The Landscape Below Ground

Proceedings of an International Workshop on Tree Root Development in Urban Soils

Edited by:
Dr. Gary W. Watson
The Morton Arboretum, Lisle, Illinois
and
Dr. Dan Neely
Editor, International Society of Arboriculture
Scott City, Missouri

Presented by The Morton Arboretum
Held September 30 and October 1, 1993

Published by
International Society of Arboriculture
P.O. Box GG
Savoy, Illinois 61874 USA

Printed by: United Graphics, Inc., Mattoon, IL

The International Society of Arboriculture
Web site: http://www.isa-arbor.com
Email: isa@isa-arbor.com

10 9 8 7 6 5 4 3 2

ISBN 1-881956-06-7

PC-1000-10/99

Preface

Many of us share an admirable goal of planting, nurturing, and preserving trees. Trees and other woody plants are fundamental to our horticultural landscapes and natural ecosystems, and to the functioning of a healthy global environment. Still, our good intentions and best efforts are not always successful. We are challenged by the stressful and altered environments in which we often want to plant and maintain trees. Many of the conditions that we seek to improve in our landscapes by planting trees are the same conditions that make tree establishment and growth so difficult.

The Morton Arboretum developed the *Landscape Below Ground* workshop to address one of the primary factors that relates to the success of tree plantings in the landscape: the root-soil environment. The Morton Arboretum is dedicated to the practical study of trees and their environments, including their planting and care in stressful, urban and suburban landscapes. One of the emphases of the Arboretum's Urban Vegetation Laboratory, part of our broader Research Program, is the study of the underground environment of trees, especially root system-soil interrelationships.

We were especially proud to arrange this great gathering of some of the top research scientists in the field, to provide them with an opportunity to interact and discuss their research, and then to share their accumulated knowledge with professionals and others interested in these topics.

Through practical research, high-quality collections of woody plants from throughout the world, and educational programs such as the *Landscape Below Ground* workshop, The Morton Arboretum strives to make meaningful contributions to the field of arboriculture. Our goal is to assure that more trees will be planted and thrive as a result of our work and this conference, endowing future generations with a greener, healthier, and more beautiful world.

Gerard T. Donnelly, Ph.D.
Director
The Morton Arboretum

These proceedings from The Morton Arboretum's *Landscape Below Ground* workshop were produced jointly with the International Society of Arboriculture (ISA). The Morton Arboretum and the ISA are both dedicated to making leading contributions to the understanding of trees and arboriculture. This partnership allows us to achieve more working together than we could accomplish working independently.

The Morton Arboretum was founded by Joy Morton in 1922 as "A not-for-profit educational institution for practical, scientific research work in horticulture and agriculture, particularly in the growth and culture of trees, shrubs, and vines by means of a great outdoor museum arranged for convenient study of every species, variety, and hybrid of the woody plants of the world able to support the climate of Illinois...to increase the general knowledge and love of trees and shrubs, and to bring about an increase and improvement in their growth and culture."

Internationally renowned as a research institution, The Morton Arboretum is directed by a Board of Trustees consisting of Charles C. Haffner, III, Chairman, Elisabeth G. Bacon, Laurence A. Carton, Natalie Culley, James F. Dickerson, Neil McKay, Henry B. Pearsall, Joseph E. Rich, and Gerard T. Donnelly, Ph.D., Director.

The International Society of Arboriculture is a scientific organization devoted to the dissemination of knowledge in the care and preservation of shade and ornamental trees. It was founded in 1924 by a group of 40 persons engaged in some phase of shade tree work or research. The group was called together in the City of Stamford by the Connecticut Tree Protection Examining Board to discuss shade tree problems and their possible solutions.

Today the Society has grown to over 8,000 members divided into twenty-seven regional and state chapters. The Canadian and European communities are now populated with four and five chapters respectively. Five Professional Affiliations consist of the Utility Arborist Association, the Municipal Arborists/Urban Foresters Society, the Arboricultural Research and Education Academy, the Society of Commercial Arboriculture and the Student Society of Arboriculture.

TABLE OF CONTENTS

PART I — TREE PLANTING AND ESTABLISHMENT

PART II — MANAGING ROOT SYSTEMS AND SOIL ENVIRONMENTS OF ESTABLISHED TREES

INTRODUCTION

It has been said that 80 percent of all landscape tree problems start below ground. Yet the below ground parts of the tree are the most difficult, and often the most expensive to study. Study often requires the destruction of the specimens. Despite these challenges, our base of knowledge continues to grow, due in great part to the scientists who participated in the *Landscape Below Ground* workshop.

Concerns about root system development of landscape trees should begin at propagation. Containerized plants often develop circling roots that can lead to serious problems when they are planted out in the landscape. Use of properly designed containers, and proper treatment of the root ball at planting time, can help to prevent circling roots.

Transplanting is an unnatural process. Over 90 percent of the root system can be lost. During the time that the root system is regenerating, water stress can develop very quickly. The time required for the tree to establish on its new site is affected by many factors, and may take years in northern climates. Severing roots during the transplanting process can lead to girdling roots.

Experienced horticulturists will sometimes spend more on preparing the planting site than on the plant. Alkalinity, deicing salts, and oxygen depleted rooting environments create unseen stresses for trees. These stresses particularly affect the growth of trees in planters. The challenge is to provide a suitable site that is also economical.

Providing specially designed root space under pavements has been successful on some sites, but sometimes roots growing under pavements are considered a nuisance. Physical and chemical barriers are being developed, but sometimes installing them around existing trees can cause major root damage. A study on root damage from trenching produced some unexpected results.

Soil compaction and resulting poor aeration is one of the biggest problems faced in the landscape. Measuring compaction and aeration can also be a challenge. Incorporation of light-weight aggregates before plants are installed can provide long-term benefits, but remedial action is often the only choice. A variety of equipment is available for this job.

The most current information available on all of these subjects is included in these proceedings. The extensive interaction among the researchers that took place during these few days will certainly generate ideas for new research that can be reported at the next *Landscape Below Ground* workshop.

Gary W. Watson
The Morton Arboretum
Workshop Organizer

ACKNOWLEDGEMENTS

The *Landscape Below Ground* workshop was a unique opportunity for many of the top researchers working with urban tree roots and soils to come together and exchange ideas, to present their latest research findings to a receptive audience, and to produce a video. Gathering such a large group of speakers together could not have been possible without substantial grant support. The following organizations provided support to help make the workshop successful:

- Arboricultural Research and Education Academy
- The F.A. Bartlett Tree Expert Co.
- Hendricksen, the Care of Trees
- Illinois Department of Conservation
- Illinois Landscape Contractors Association
- Illinois Nurserymen's Association
- USDA Forest Service, Urban Forestry Center for the Midwestern States
- USDA Forest Service, Northeastern Area State and Private Forestry

The gratitude of The Morton Arboretum and all of the presenters is also extended to the International Society of Arboriculture for agreeing to publish these proceedings.

PART I

TREE PLANTING AND ESTABLISHMENT

Size, Design, and Management of Tree Planting Sites

Thomas O. Perry

The article describes simple methods for examining tree roots: where tree roots grow in both typical and atypical soils and why; what plants must obtain from the soil in order to prosper; and the minimum land area, soil depth, and water requirements for trees of different sizes. Finally, a brief list of other conditions required for vigorous growth and long life of urban trees is provided.

Typical Tree Roots and How to Examine Them

The four to eleven large roots that radiate from the base of the tree trunk ("the root collar") are usually easy to see. They are either on the surface of the soil or only a few inches below. They can be very large as they originate from the root collar of a typical tree, but they taper rapidly and branch repeatedly so that at distances of 10 ft or more from the trunk they are usually about the size of your thumb or smaller. They grow horizontally through the soil and commonly extend for 40 feet or more from the branch tips (Figure 1). They remind one of ropes as they extend for amazing distances through the soil. In the forest, roots from different trees crisscross each other and form a complex network. Sometimes the roots of one tree extend next to or under the stumps of neighboring trees. Frequently, roots of several trees of the same species are grafted together.

All of this can be demonstrated with a half hour of industrious digging with a sharp stick or an ordinary trowel. Some of the labor of digging can be avoided by examining new highway cuts and real-estate developments or visiting areas subjected to a hurricane, tornado, or heavy winds after soaking rains (Fig. 2). Almost any walk in the woods will reveal the pancake of roots and soil that is heaved up when healthy trees topple over.

These larger roots serve multiple functions. They transport the water and minerals absorbed by the smaller roots of the tree. Like the smaller roots, they contain cells that synthesize a variety of substances essential to the normal functioning of the top of the tree. Finally, they function to hold the tree up. For simplicity, let us call these larger roots "transport roots."

Most tree roots range in diameter from that of the lead in a pencil to the size of a hair. These smaller roots originate from the larger roots described above and grow upward into the surface inches of soil and the litter layer. Yes! Most tree roots grow up — not down! These smaller roots branch four or more times as they grow upward to form mats in the surface and litter layers of the soil. These smaller roots are often nonwoody (16)

Thomas O. Perry is Professor emeritus, Forestry and Landscape Architecture: NC State University, 5048 Avent Ferry Road, Raleigh, NC 27606

Figure 1 — Root of an oak tree exposed by digging with a sharp stick and then painting white to contrast with the soil background. This transport root originated from one of the main roots at the root collar of the tree in the background. Work of Walter Lyford. Harvard Forest.

and die and reform several times during the year. There can be hundreds of roots in a cubic inch of soil (Fig. 3). Small ephemeral roots grow quickly up into the leaf litter and die back into the A-horizon with equal speed after a drought or freeze. These smaller roots with their huge surface area and intimate contact with the soil function primarily in the absorption of water and minerals from the soil. They are also sites of synthesis of asparagine and glutamine and other amino acids. For simplicity, let's call these "absorbing roots".

The surface area of the absorbing root system of trees is normally associated with fungi. These fungi obtain energy and essential organic matter from the tree. The tree benefits from the extra water and mineral absorption capacity provided by the vast surface area of the semi-microscopic strands of the fungi which extend outward from the roots for several centimeters. The complex of the root and the fungi is referred to a mychorrhizal root.

The first root of many tree species (especially those with large seeds) may grow down into the soil for several feet to form a tap root. Other "sinker" roots may originate at various points along the main root system of the tree. These tap and sinker roots grow down only into the layers of soil where there is sufficient oxygen. This may be only several inches in heavy clay soils or in areas with a high water table or it may be several feet in soils with lots of pore space for easy penetration of oxygen. A secondary set of transport roots usually grows horizontally from these tap and sinker roots. Feeder roots usually grow upward from these lower layers of transport roots. In many instances, when dense layers of soil with little or no oxygen are encountered, the sinker and tap root system stops and abruptly turns 90° (Fig. 4).

Figure 2 — White oak root system. This is the characteristic pancake of roots and soil heaved up when soaking rains plus winds combine to blow trees over. This shallow root system is characteristic of the form developed in typical soils in the world.

Roots have important functions other than absorption of water and minerals:

- They are the sites of synthesis of amino acids, cytokinins, and other organic compounds that are transported up through the xylem of the tree to the leaves. These serve as raw materials for synthesis of proteins, chlorophyll, and other molecules essential for plant growth.
- They act as cables and networks that bind the soil and hold the tree up. In strong winds the tree is a vertical cantilever with the roots and mass of the soil being the support system for the swaying tree top. The roots on the windward side of the tree are placed under tension and the roots on the leeward side are placed under compression. When the winds are too strong things break; sometimes limbs and trunks and sometimes roots. Sometimes a huge pancake of roots and soil is heaved out of the ground as the tree topples over. This is particularly likely to happen when the soils are softened by large amounts of water associated with storms.

Figure 3 — Photograph of roots exposed by carefully working with ones fingers, a sharp stick, and a whisk broom. Roots sprayed with paint for contrast with the soil background. Feeder roots in front of the trowel were in the litter layer and upper inch of soil. Transport roots in back of the trowel were exposed by removing the feeder roots and the upper inches of soil. Feeder roots grow up from the transport roots.

All roots function as absorption and transportation roots and the variation in size of roots is continuous. Hence the distinction between absorbing roots and transport roots is not as clear cut as described in the preceding paragraphs. However, all studies indicate that the uptake of water and minerals takes place primarily through the small diameter, non-woody absorbing (or feeder) roots that are concentrated in the uppermost inches of soil.

Roots are opportunistic and grow anywhere that the resources of life are available: in the air of super humid rain forests; down through the cracks in rocks into Mammoth Cave, Kentucky; into the moisture laden crevices of mountain tops; down storm sewers; and under sidewalks.

Trees will heave pavement! When the earth beneath pavement is well aerated, and a moist layer of sand and gravel underlies the pavement, trees should not be planted nearby unless a barrier to their growth is provided.

It is very easy to stop the growth of roots. They cannot penetrate soils that have limited oxygen supplies, heavy clay contents, and are well compacted. They will not grow where there is no oxygen. Neither will they penetrate well constructed concrete barriers or carefully installed plumbing that have no cracks or crevices. Phil Barker's research with the U.S. Forest Service shows that even a thin layer of 6 mil plastic is sufficient to stop penetration by tree roots (1). Carl Whitcomb's research with porous plastic shows

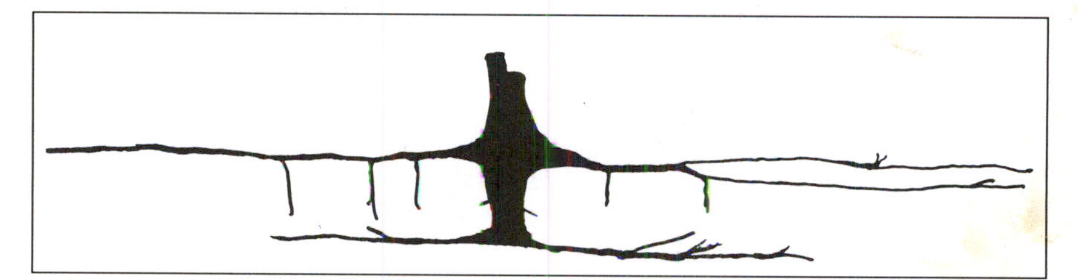

Figure 4 — In sandy soils or in soils where there is oxygen at greater depths many trees can develop a taproot and sinker roots plus a second layer of horizontal roots. The second layer of roots forms at the interface between those soil layers that bear oxygen and those that do not. Root systems of this type are common in the sandy soils of the Southeast Coastal Plain.

that roots are inhibited effectively by tightly woven plastic even though water and oxygen are freely available on both sides of the barrier (19).

The Water Required for Plant Growth

Many researchers have estimated the rates of water loss from forested and agricultural ecosystems. Rates of water consumption are closely correlated with the flux of radiant energy and the available soil moisture. The studies of Metz and Douglas (17) have been combined with the previous studies of spacing requirements for trees of various sizes to yield Table 1. The selection of rates of water consumption of 0.1, 0.25, and 1.0 acre-inch per day for the attached tables requires justification:

The flux of radiant energy on a sunny July day can be between 500 and 600 calories per square centimeter. This translates to a maximum potential evapotranspiration of 0.25 acre-inch per day. This rate of water consumption can occur only when there is an ample supply of soil moisture. The easily available soil moisture will be depleted after several days of high radiant energy flux, and actual evapotranspiration will decrease. Evapotranspiration rates of 0.1 acre-inch per day are characteristic of conditions of moderate water stress. When radiant energy and winds are focused by reflecting pavement and buildings, evapotranspiration rates can be much greater than those measured in a standard evapotranspiration pan—as much as 1 acre-inch per day.

Plants may wilt because there is too much water and the oxygen and gas exchange processes in the soil are blocked. Plants may wilt because of soil compaction or because they have diseases that destroy roots or destroy the osmotic integrity of cell membranes. These alternative causes of wilt or stress should be checked before watering plants. Too much water can be as harmful as too little water.

Do not overwater! A ten-inch layer of typical soil can hold about 1 inch of water for plant use. Watering beyond this amount is very likely to do more harm than good. In many environments about 1 inch of water per 7 to 10 days should be sufficient. Millions of dollars worth of landscape plants are killed every year by overwatering and malfunctioning irrigation systems. The waste of water makes this a double tragedy.

Loss of 1/4 inch of water per day amounts to 156 gallons per 1000 sq ft per day or 1090 gallons per week (~ one gallon of water per sq ft of land per week without rain). Typical soils will not allow penetration of this amount of water in a single dose. Two or three applications are preferable to a single large application. No! The old saying "water deep so the roots will go deep" is not correct. Roots will penetrate only the layers

Table 1. Gallons of water consumed per tree per day as calculated from the amount of water (acre-inches) consumed by a fully-stocked acre of trees of a particular dbh. For example: A healthy 5" dbh open-grown hardwood will require 7.4 gallons of water per day with moderate water stress; 18.6 gallons per day without stress in a normal environment; 74.4 gallons per day without stress in an extreme urban environment. In practice, the gallons need not match the false precision of the calculated numbers in the table—two digit accuracy will do

	Forest-Grown Pine			Forest-Grown Hardwood			Open-Grown Hardwood		
DBH	acre-inches			acre-inches			acre-inches		
inches	0.10	0.25	1.00	0.10	0.25	1.00	0.10	0.25	1.00
1	0.3	0.9	3.5	0.5	1.2	4.9	1.2	3.1	12.3
2	0.9	2.4	9.5	1.3	3.2	12.8	2.3	5.7	22.9
5	3.5	8.8	35.2	4.6	11.5	45.9	7.4	18.6	74.4
10	9.5	23.8	95.2	12.1	30.2	120.7	22.6	56.4	225.7
15	17.0	42.6	170.3	21.2	53.1	212.4	45.9	114.7	459.9
20	25.7	64.3	257.3	31.7	79.3	317.2	77.4	193.5	774.1
25	35.4	88.6	354.4	43.3	108.3	433.0	117.1	292.8	1171.1
30	46.1	115.1	460.4	55.8	139.6	558.3	165.0	412.5	1650.0

of soil where there is adequate pore space. Both home owners and landscape managers can use pie plates to gauge the amount of water applied. One third of an inch of water in the pie plate should signal enough for the day! It is time to shut off the irrigation when the earthworms come up for air and the robins come around to feast on them.

The Soil Required for Plant Growth

As described previously, the growth rates and quantities of living trees and plants were correlated with a multitude of variables. However, within a given region, the variation in the vigor, growth rate, survival, maximum biomass, leaf area index, etc. of trees are influenced primarily by properties of soil that are correlated with oxygen, moisture, and nutrient availability. These include:

1. Soils with continuous pores that allow easy penetration of oxygen, water, and tree roots are the most productive.

2. Soils that retain water and minerals in forms available to roots are the most productive.

3. Soils that are ideal for tree growth are about 2 to 3 % organic matter (by weight) and have a blend of sand, clay, and silt which provides a good compromise between oxygen and root penetration and water and mineral retaining capacity—a "sandy loam". Soils with an excess of clay or other fine textured material are poor supporters of plant growth.

Much of the critical pore space for oxygen and root penetration is provided by nature's plowmen — earthworms, crickets, moles, mites, and thousands of other creatures that use the organic matter from leaf litter, decaying roots, and other plant parts as energy sources for their tunneling activities. Surprisingly, although the organic matter content of soils is only 1 to 2 % of the total mass, it equals nearly 1/2 as much as the organic matter present in living trees and plants on the land (7).

4. Soils that are thin, because bedrock or water tables are close to the surface, are less productive than soils that are about 30" thick. Water and minerals are quickly lost from soils that are too coarse in texture or too deep
—poor tree and shrub growth can be sustained on soils that are as thin as 5 inches.
—fair tree growth can be sustained on soils that are as thin as 10 inches.
—good tree growth can be sustained on soils that are 16 inches thick.
—excellent tree growth can be sustained on soils that are 20 to 30 inches thick.
—tree vigor decreases gradually as soils increase in thickness beyond 30 inches.

The work of Coile (4) and his students (2) illustrate the relationships between soil depth, soil texture, and plant vigor (Figure 6). Although the curves will vary, the pattern of these relationships hold for all plant species. Craul's book "Urban Soil in Landscape Design" (5) provides a definitive review of the literature on the relationship between soil properties and tree growth.

The properties of natural soil change with depth. The surface inches of soil are usually coarser and texture, have better structure, and contain the most organic matter and available nutrients. They also provide the conditions most favorable for plant growth (Fig. 5).

Publications and research by Lindsey (13,14), Lindsey and Bassuk (15), Goldstein, et. al. (8), and Urban (20), report the planting site requirements of trees in terms of cubic feet of soil and soil water availability. When a soil depth of 2 feet is assumed their data for smaller trees agree remarkably well with older research and the data for forestry (Fig. 6). The publications of Kopinga (11) also fall within the limits indicated in Fig. 6. The agreement would be even better if Urban recognized that the area required for tree growth increases in proportion to the square of the diameter.

Horizontal space, water, and soil volume are only three of the variables to consider in designing tree planting sites. The area of soil exposed to the atmosphere is important! Plants will grow better in 100 cubic feet of soil with a 10 by 10 foot surface area than they will in 100 cubic feet of soil with a 1 by 1 surface area. Roots play a vital role in supporting a tree and room must be provided for their outward extension if a tree is to stay upright in strong winds. Other requirements for successful urban-tree plantings are described in the following sections.

What Plants Must Obtain from Soil

In order to grow vigorously and live a long time plants must obtain the following things from the soil (Fig. 7). Each is equally important. Failure to provide any one of them will result in failure of the planting.

Water

The quality and quantity of water must be carefully regulated. Precipitation must equal or exceed evaporation and excess salts and noxious substances must leach away. Too much water, even briefly, produces anoxia and root death. Plants will wilt both when there is too much and too little water. Nearly all urban planting sites will require drainage for removal of excess water and for the flushing of excess salts.

Oxygen

Roots require oxygen—no oxygen, no plant roots! Plants genetically adapted to river

Figure 5 — Growth of yellow birch seedlings (Betula lutea Mich.) in different layers of a forest soil. The pot on the left (H) contained soil from the top 2 inches of soil. The next pot (A) contained forest soil from a depth of 2 to 4 inches—in the A-horizon. The B pots contained soils from the B-horizon. The pot on the right (C) contained soil from the C-horizon. This study by Hoyle (10) demonstrates dramatically the importance of protecting and saving the uppermost inches of soil. Jumbling and respreading the top foot of soil from a construction site is a commendable gesture. Separating the top inches and placing them back on top after construction is complete would be even more commendable.

bottoms and swamps possess special air passages and biochemistries that allow some penetration into anaerobic soil.

Support

Too many trees fall over when they are planted in urban situations that do not allow normal cable support by radial extension of tree roots. Trees in "potting soil" mixtures that include as much as 1/3 of their volume in organic matter tend to fall over as the oganic matter decays. Support can be especially important in cities like Boston, New York, and Chicago where tall buildings tend to fetch gale force winds down to street levels.

Warmth

Roots do not go dormant and may grow 12 months of the year. Tree roots die when soil temperatures fall below about -6° C (24°F). Forest soils are covered with an insulating blanket of leaves and rarely freeze even in northern Maine and Minnesota. Unlike roots, tree tops, when dormant, can survive temperatures as low as -30° C (3,6,9,12,18).

Nutrients

The relative concentrations must be appropriate and the pH must be right for the species in question. Excess of one ion can block the proper uptake of another.

Stability

The soil and its environment must be stable. This means that the site must be large relative to the size of the plant. The soils must not be shifting like desert sands or heaving as a result of frequent frosts. The soils must be thick enough and sufficiently well drained to

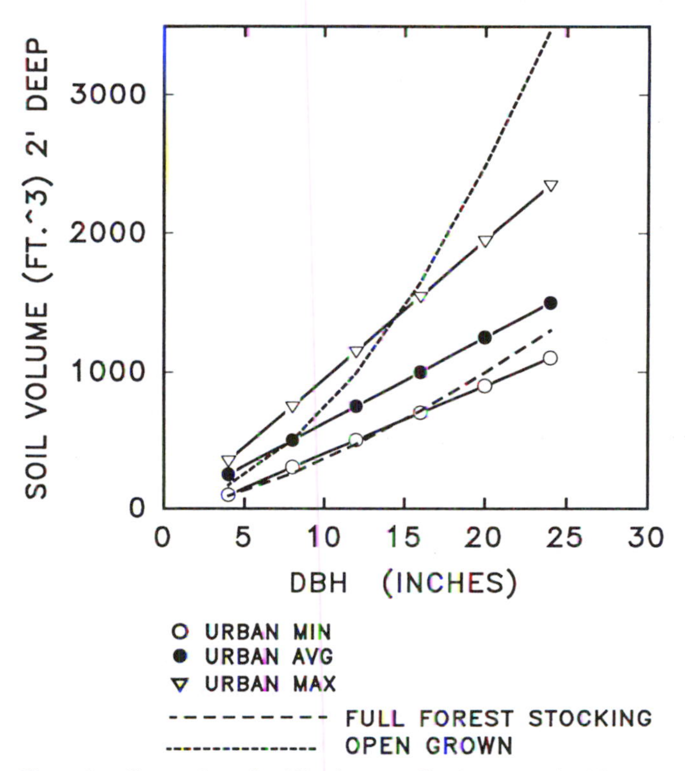

Figure 6 — Comparison of publications on soil volumes required for tree growth. For smaller trees, the agreement among researchers on the minimum volumes required is remarkably similar and would be even closer if everyone recognized that the resource requirements of trees is an exponential function of their diameters. Large open-grown trees require much more soil and space than indicated by the straight-line extrapolations of Urban's data. There are other variables to consider.

prevent rapid fluctuations in temperature, oxygen, or water content. Horticulturists must not be constantly tearing up the soil (and tree roots) to plant new shrubs or herbs.

Freedom from Toxicants

Deicing salts, herbicides, old paint residues, scraps of sheet rock, cement, etc—nematodes, and pathogens.

Room to Grow!

Roots compete with each other just as the leaves in the tops do. The availability of water, oxygen, and minerals for the roots can limit plant growth just as completely as the availability of light for the tops

Other Conditions Required for Plant Growth

More than 20 years of repeated visitation and photographic recording of planting sites throughout the U.S., Canada, and Great Britain reveals that, in addition to the soil

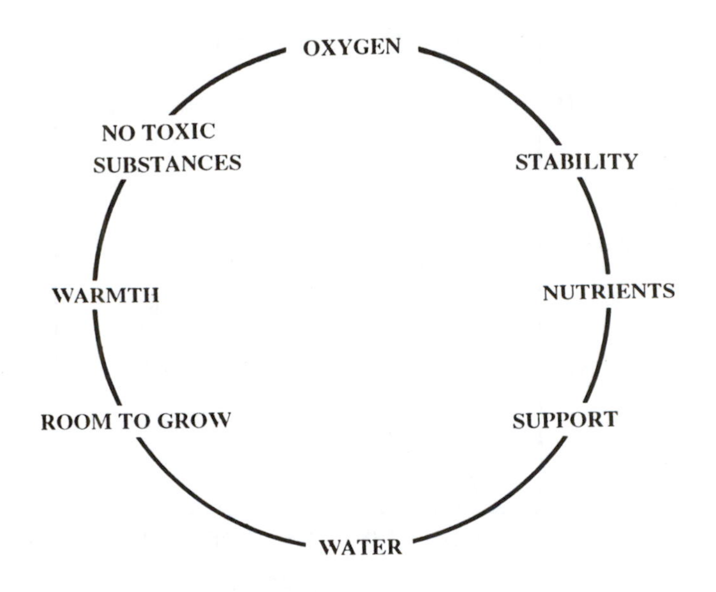

THINGS PLANTS MUST OBTAIN FROM THE SOIL

Figure 7 — List of things plants must obtain from the soil if they are to grow rapidly, live a long time, and be attractive. The list is arranged in a circle because lack of any one of the things will lead to premature failure.

requirements listed above, the following must be provided in a long term landscape planting:

1. Room and opportunity for the top to grow. Adequate crown space including both lateral and overhead clearance, and clearance for pedestrians and tall vehicles.

2. Adequate light. Lighting can be a problem in the skyscraper canyons of today's cities. The east and north sides of streets can be reflector ovens while the south and west sides of streets may not receive sufficient light except when the sun is high in the sky.

3. Protection from scuffling feet, leaning bicycles, weed eaters, automobiles, bumps due to loading and off-loading goods to local shops, and general vandalism. Protection from soil compaction by people, pigeons, and maintenance or other vehicles. Protection from excess populations of squirrels, pigeons, insects, and other pests.

4. Shelter from focused winds and the reflector-ovens created by reflections of buildings and pavement. Landscape plants can be cooked in one day when placed on the south-west side of a glass building which is surrounded by pavement!

5. Clean air! If the air is not good for humans, it is not good for plants. Direct exposure to rising exhaust fumes from buses, cars, automobiles and chiminies is as lethal to plants as it is to humans.

6. Provision for routine maintenance and inspections, repair and inspection of irriga-

tion, soil sampling, removal of trash, unplugging of drains, provision of winter mulch, replenishing of depleted organic matter. Heavy tree grates and hollow spaces under the pavement can present special problems. Make sure these heavy items can be removed without undue labor and expense.

7. Avoidance of heaving pavement by roots. This may require root barriers or conversely require tree grates or hollow places under the pavement.

8. Design for thinning and pruning! Even at 50-foot spacings, healthy attractive trees will grow so that their crowns will interfere with each other, adjoining buildings, passing traffic, and pedestrians. Arrange the pattern of planting so that trees left after thinning can reoccupy the site and provide a pleasing effect. Anticipate patterns of growth that will block critical vistas.

9. Design for replacement of either plants or soil or both. In many urban situations trees will outgrow their assigned space or simply loose vigor and attractiveness. The landscape design and its infrastructure should allow for easy replacement of plants that have outlived their usefulness. Soils can become hopelessly compacted and loaded with deicing salt and other toxicants. The number of instances when the planting has to be dug up and the soil thrown away is increasing.

10. Money for maintenance, thinning, and replacement. Nothing is permanent — especially living trees struggling to survive in stressful urban environments. Maintenance of urban tree plantings will be much more expensive than comparable plantings in open parks and forests. Half dead trees in unmaintained planters look terrible. Don't plant trees unless there are adequate funds for proper maintenance.

11. Allowance for road repairs, utility repairs (including irrigation!) and future utility installations (including fiber optics and other underground utilities).

Planting Site for a Three Inch Caliper Tree

All of the preceding data and observations can be combined to prepare specifications for planting a tree with room for it to grow:

If healthy growth of the tree is desired, roots should be encouraged to grow beyond the planting hole into the site as rapidly as possible. There has been too much focus on the "hole" and not enough on the site as we plant trees.

The accompanying drawing, including the tree is to scale (Fig. 8). Beware of architectural details that do not show proper scale for both the tree and the surrounding soil. If possible, the bottom of the root ball should be seated on undisturbed solid soil. For lateral support, the lower half of the root ball can be surrounded by well compacted soil. This is to keep the tree from shifting and settling. Ninety percent of the roots will grow out of the upper portion of the root ball and into the upper layer of good soil.

With sufficient room and good soil, trees can grow 3/4 inch or more in diameter each year for the first 20 years after they are planted. It is not uncommon for 30-year-old trees to achieve diameters of 20 inches in good soils.

There are problems of water and oxygen movement between soils of different structure and texture and corresponding problems of root growth and development. Whenever possible, the soil of the root ball and the soil of the back fill and the soil where the roots are to grow should be as similar as possible. Use the same soil for the backfill as the native soil. The backfill soil should not contain more than about 2 to 3% organic matter by weight.

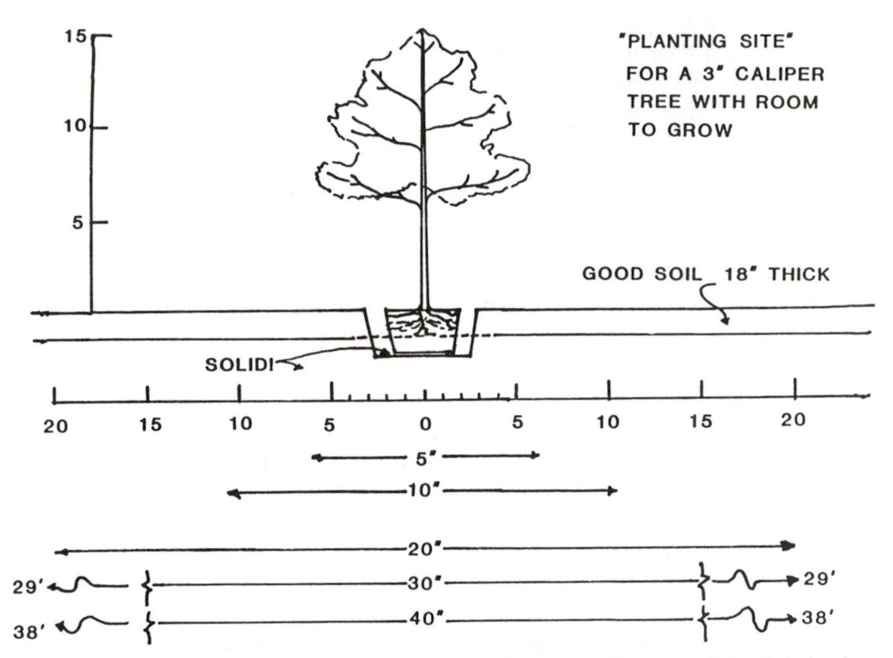

Figure 8 — Diagram of a planting site for a 3-inch-caliper tree. The tree and the drawing is to scale. The arrows below the drawing are based on the data for open-grown trees in Table 2. The arrows indicate the space that ideally should be prepared at the time of planting and kept well watered and free of weeds and competition. Tree roots can grow laterally 10 ft. or more per year.

Roots grow very rapidly in good, competition-free soil — as much as 10 to 15 feet per year in well prepared soil. Tree stakes should be only temporary devices which can be removed at the end of the first growing season.

The arrows and lines below the basic sketch in Fig. 8 indicate the distances to which soil should be deep plowed and tilled and kept free of grass and other competing vegetation if the tree is to grow to the size indicated and have a healthy vigorous crown. For example, if the owner desires his 3" tree to grow quickly into a 10" tree he should prepare the soil in an area approximately 20 x 20 feet.

Mulch is desirable. However, mulch is no substitute for breaking up the soil and having good soil in the entire area where roots are to grow. The mulch should not be over 3 in thick after settling— otherwise fermentation and low oxygen concentration is a hazard to the tree or shrub. Green mulch can be used. However, extra fertilizer will be required to compensate for the minerals bound by the microorganisms. Twigs and chips are superior to bark mulch. They bind together better, do not wash away as easily, and are more rapidly digested by microorganisms. The growth of tree roots and weeds in the mulch can be prevented by underlaying the mulch with porous polypropylene matting.

Mulch should not be applied within 6 - 8 in of the tree trunk. Mice like to nest in mulch during the winter and will gnaw on the bark of young trees and kill them.

Urban planting sites are usually small and unavoidably harsh. The smaller the site, the shorter will be the life of the plants therein and the more expensive will be the maintenance costs. As noted earlier, for many planting sites in downtown business districts, we are really dealing with large Bonsai plants. All designers and managers of urban tree planting sites will do well to read and memorize books on care of Bonsai plants. Note the frequency with which Bonsai specialists must dig up their plants, prune the roots, and replace the salt contaminated and depleted soil. We are presently unable to design small planting sites that do not require comparable care.

Literature Cited

1. Barker, P. 1989. Tree roots and sidewalk conflicts. AFA Proceedings. Fourth Urban Forestry Conference. St. Louis.
2. Barnes, R.L., and C.W. Ralston. 1955. Soil factors related to growth and yield of slash pine plantations. University of FL., Ag. Exp. Sta. Bull. 559.
3. Beattie, David J. ed. 1986. Principles, practices and comparative cost of overwintering container grown landscape plants. Pennsylvania State Univ. Ag. Exp. Sta. University Park, PA. Southern Cooperative Series Bulletin #313, 32p.
4. Coile, T.S. 1937. Soil and the growth of forests. Advances in Agronomy 4:329-398.
5. Craul, P.J. 1992. Urban Soil in Landscape Design. John Wiley and Sons, N.Y 396 pp.
6. Crider, F. J. 1928. Winter root growth in plants. Science 68:403-404.
7. Divigneaud, P., and Denaeyer-De Smet. 1970. Biological cycling of minerals in forest ecosystems. In Reichle, D., Analysis of Temperate Forest Ecosystems. Academic Press. pp. 199-225.
8. Goldstein, Jan, Nina Bassuk, Patricia Lindsey, and J. Urban. 1991. From the ground down. Landscape Architecture 81(1):66-68.
9. Hämmerle, J. 1901. Über die Periodizitat des Wurzelwachstums von Acer pseudoplatanus L. Beitrage zur Wissensaftliche Bot. 4:105-155.
10. Hoyle, M.C. 1965. Growth of yellow birch in a poolzol soil. Northeast Forest Exp. Sta. Research Paper NE-38.
11. Kopinga, J. 1985. Research on street tree planting practices in the Netherlands. Proceedings Fifth Conference of the Metropolitan Tree Improvement Alliance. Pennsylvania State University, University Park, PA
12. Ladefoged, K. 1939. Untersuchungen über die Periodisitat im Ausbruch und Langenwuchstum der Wurzeln. Det Forstlige Forsøgsvaesn i Danmark. XVI p255.
13. Lindsey, Patricia A. 1990. Difference in water use rates between four broadleafed woody tree species. M.S. thesis Cornell University.
14. Lindsey, Patricia A. 1993. Determining an adequqte soil volume, improving the rooting Environment, and Measuring the Water Use of Urban Trees. Ph.D. thesis Cornell Univ.
15. Lindsey, Patricia A. and Nina Bassuk. 1991. Specifying soil volumes to meet the water needs of mature urban street trees and trees in containers. J. Arboric. 17:141-148.
16. Lyford, W. H. 1975. Rhizography of nonwoody roots of trees in the forest floor. p177 In Torrey, J. G., and D.T. Clarkson, eds. The Development and Function of Roots. The Third Cabot Symposium. Academic Press.
17. Metz, L.J., and J.E. Douglas. 1959. Soil moisture depletion under several Piedmont cover types. Tech. Bull. 1206. 23 pp. USDA Forest Service.
18. Perry, T.O. 1972. Dormancy of trees in winter. Science 121: 29-36.
19. Shinozaki K. and T. Kira. 1956. Intraspecific competition among higher plants. VII Logistic theory of the C-D effect. J. Institute of Polytechnics. Series D: 7:35 —72 (in English)
20. Urban, J. 1992. Bringing order to the technical dysfunction within the urban forest. J. Arboric. 18:85-90.

Plant Establishment and Root Growth Research at the Urban Horticulture Institute

J. Roger Harris, Nina L. Bassuk and Susan D. Day

The Urban Horticulture Institute was founded as part of the Department of Floriculture and Ornamental Horticulture of Cornell University, Ithaca, New York, in 1980 because of increasing concern over the livability of our urban environments. The Urban Horticulture Institute (UHI), under the program leadership of Dr. Nina Bassuk, conducts research, disseminates information through the New York State Cooperative Extension system and coordinates workshops. In addition to Dr. Bassuk, the UHI consists of Dr. Tom Whitlow and eight graduate students under their direction.

Research at the UHI mainly concerns plant establishment in urban and community areas. Our philosophy of plant establishment is diagramed in Figure 1. The first step in establishing plants is site assessment. Planting sites have little resemblance to the adjacent woodland or field since the soil has been disturbed and the above-ground environment has altered. A detailed on-site evaluation of soil conditions (pH, compaction, texture, etc.), radiation patterns (reflected heat, shade, etc.), pedestrian patterns, deicing salt damage potential, etc. is needed at the start. The next step in the plant establishment process can be planned only after growing conditions have been evaluated.

Poor growing conditions can often be alleviated somewhat by site modification techniques. For example, poor drainage can often be overcome by the installation of subsurface drainage systems or the redirection of surface water. Compacted soil can often be replaced with better soil, or subsoiling may prove to be effective. The installation of curbing may help alleviate the harmful effects of deicing salts. Unfortunately, the ability of site modification to improve growing conditions often limited because of the nature of the site or budget constraints. Alteration of pH, for example, is not a good permanent solution for long-lived plants with far reaching root systems such as trees. Subsoiling is really only effective if you can break through to freely draining soil, and it can be very destructive to the root systems of existing plants. Subsurface drainage may be impractical, and it only removes water moving by gravitational forces. Fine textured soils may remain too wet for many species, therefore, even when the drainage system is installed. In these cases, a redesign of the project and the selection of tolerant species is appropriate. For example, the grouping of plant material in common planting areas promotes a more efficient use of the available soil volume than if plants are confined in individual

---

J. Roger Harris, Nina L. Bassuk, Susan D. Day, Jason Grabrosky, Tony Dufour, Anna Perkins, Lou Anella, and Barbara Neal are with the Urban Horticulture Institute, 20 Plant Science, Cornell University, Ithaca, NY 14853.

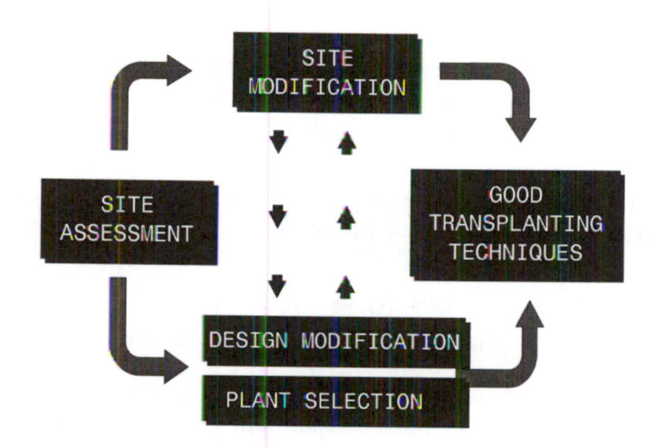

Figure 1 — UHI method for plant establishment

planting pits. In addition, a 'forest effect' can be achieved when plants are grouped, creating a more favorable above-ground environment. Planting areas can be moved away from the high pH found around buildings and sidewalks.

The selection of plant material that will tolerate specific conditions after final modifications and design alterations are made is essential. In other words, one should only choose plant material which will tolerate the specific site conditions. Selection for aesthetic qualities should only be made from that list of plants. Finally, good planting techniques are essential, but only after a detailed site assessment has been made, feasible site modifications have been implemented and the selection of species which will tolerate the final conditions compiled. Research at the UHI involves all phases of this process.

Some of the past research projects have included research on the physiological basis of drought resistance in trees (5), the description of the urban environment and associated stresses on street trees (7), the propagation of desirable urban tree species (4,6), the development of an easily implemented street tree inventory method (2) and the determination of soil volume needs for street trees (3). Some of the current research projects are summarized below.

Aggregate Mixes for Use Under Pavement

Jason Grabrosky, Graduate Student

Many of the problems associated with the decline of street trees can be attributed to inadequate soil volumes. Tree roots are seldom able to penetrate beneath the streets and sidewalks of our downtown environments because of the compaction levels required for the soil underneath pavement. Tree roots are instead confined to usually no more than a 4 X 4 X 3 = 48 ft^3 volume of soil. A medium-sized shade tree in reasonably good soil would instead require 300 - 400 ft^3 of soil without additional irrigation in the northeastern United States (3). A tree whose root system approaches fully exploiting the 48 $ft.^3$ volume of soil found in a typical tree pit would be increasingly vulnerable to attack by insects, diseases and people pressure and therefore increasingly susceptible to decline. The intent of this research is to develop a soil mix that will meet engineering standards for compaction yet retain characteristics which will be favorable for exploitation by tree roots. Such a mix is not intended for the primary growing media, but for a 'breakout zone'. A mix that meets the engineering standards for compaction, yet provides for sufficient aeration and lack of impedance requires that the mix be a solid matrix of aggregates (such as gravel). Tests are therefore underway which tests different types of aggregates combined with different amounts and compositions of media. Engineering specifications are being developed for many combinations. The combinations which meet the engineering specifications are then being tested for plant response by compacting the mixes in containers and planting with trees. The mixes which hold the most promise will be further screened in a full-sized tree pit and pavement test.

pH Tolerance Trials

Tony Dufour, Graduate Student

Most literature on plants and pH is observational and only states what is best, or what is 'preferred' for the particular plant in question. In order to make meaningful plant selections, we instead need to know the limits of pH tolerance for tree species. The pH in most urban plantings is elevated (more alkaline) due to the leachate from mortar used in the making of concrete and cement. Trees which can tolerate these high pH levels are needed for urban plantings. The goals of this research are to 1) Develop a protocol for testing the pH tolerance of trees; and 2) Identify species or populations of species which are tolerant of high pH.

The first trees to be evaluate are eight species of Quercus (oak). Acorns are being collected this fall from trees known to be growing on elevated pH sites from many parts of the country and from trees known to be growing on sites with a pH more typical for the particular species. Acorns from these trees will be grown and tested for tolerance to high pH. The research will expand to other species when the testing protocol has been satisfactorily established

Vegetative Propagation and Production of Desirable Plants for Urban Landscapes

Anna Perkins, Graduate Student

Many species which are most resistant to urban stress or have very desirable aesthetic qualities are also very difficult to propagate vegetatively (e.g. Nyssa, Corylus, Stewartia). Selection for tolerance to urban conditions within species therefore cannot be easily accomplished. A method of cutting propagation utilizing an etiolation technique was developed by previous UHI research (4,6). The purpose of this research is to expand the number of species tested using this propagation method and to develop methods for growing the rooted cuttings through the first overwintering and up to the lining out stage. Some species propagate readily, but post-rooting survival is low. The effect of several growth regulators as well as the effect of various storage conditions are being evaluated.

The Effect of Provenance on Flooding Tolerance of Red Maple

Lou Anella, Graduate Student

Acer rubrum (red maple) has a natural range from Canada to Florida, and it can be found growing in low flooded areas as well as on dry ridge tops. The purpose of this research is to determine if a difference in tolerance to flooding between those populations (provenances) growing in flooded areas and those growing in dry areas exists. Another objective is to identify the physiological basis for such a difference, if it exists.

A better understanding of the physiology behind the ability to tolerate adverse conditions such as flooding will increase our ability to choose plant material for potentially stressful landscape situations.

One isolated population of trees growing along the bank of the James river in northern Virginia and another growing on a mountain ridge top in western Virginia were identified for testing. Seeds were collected from each provenance in 1990 and sown in a greenhouse. Seedlings were then grown in containers until testing for flood tolerance began in 1993. Trees from each provenance were flooded and photosynthesis rate, stomatal conductance, and morphological changes were recorded.

Trees from the wet provenance quickly reacted to flooding. Photosynthesis rates and stomatal conductances quickly dropped, whereas reactions were much slower in the trees native to the dry habitat. Photosynthesis rates eventually dropped to lower levels in the dry site trees, however. Abscisic acid (ABA), a plant hormone, is being investigated as a possible root-to-shoot signal on the wet site trees. In addition, the wet site trees developed raised lenticels (hypertrophy) along the lower trunks, but trees native to the dry sites did not. It is thought that hypertrophied lenticels and an accompanying increase in the air filled porosity of stems act as a conduit of oxygen to flooded roots. The anatomy of the stems are being studied to determine if wet site trees developed larger internal conducting areas after flooding.

Evaluation of Effect of Tree Pit Construction on Tree Growth

Barbara Neal, Graduate Student

Tree pits were designed and installed with ultimate tree health in mind on Pennsylvania avenue and adjacent highly visible areas in Washington, D. C. in the late 1970's. The expense of mixes and pits were of less concern than the effect on tree health because this is the street on which the presidential inaugural parade is centered. Street trees were therefore intended to become a showpiece and a source of national pride. The purpose of this research is to measure the effect of this 'first class' treatment on the trees which were planted into these areas. Incremental growth patterns will be determined by cores into the center of the trunks so that yearly growth can be assessed. Comparisons between these various designs and standard designs can then be made

Remediation Techniques for Trees in Compacted Soil

Susan Day, Graduate Student

Soil compaction is a perennial problem on most landscape sites. Compaction usually occurs because of heavy equipment traffic during construction. Soil compaction destroys soil macropores by compression, thus reducing drainage and increasing resistance to root penetration. Slow drainage decreases oxygen diffusion to roots since oxygen travels much more slowly through water than through air. This slowed oxygen diffusion limits the active respiration which is required for normal growth and health of roots. The reduced root extension caused by high penetration resistance and lower growth rates results in trees which are more susceptible to drought and mineral deficiencies.

The purpose of this research was to test several method for effectiveness of alleviation of the adverse effects of soil compaction. A field of silty clay was compacted to a bulk density of 1.5 g/cm^3 by making repeated passes with a large tractor. Two species of landscape-sized trees were then planted into an experimental design consisting of five treatments. Callery pear (*Pyrus calleryana*) was chosen for its reported tolerance to compacted soil, and sugar maple (*Acer saccharum*) was chosen for its reported sensitivity to compacted soil. All trees were approximately 2 M tall and were planted bare-root into the following treatments in the spring of 1991: 1) Backfilled with existing soil (control); 2) Backfill amended with 50% v/v spaghnum peat (amended backfill); 3) Four trenches 0.3 M deep and 1 M long filled with sandy loam soil radiating in a spoke-like pattern from the planting hole (soil trenches); 4) Four 0.6 M long X 10 cm diameter perforated ADS drain pipes installed vertically 1 M away from the trunk and equidistant from each other (vertical drains); and 5) Four drainage panels 0.3 M wide X 0.6 M long X 1.5 cm thick placed on the long edge in the ground but with contact with the surface air, installed equidistantly in a spoke-like pattern around the planting hole (Enkadrain®). The effect of the treatments on soil oxygen concentration was measured periodically, and shoot extension was measured after the 1992 and 1993 growing season. Sugar maple trees suffered high mortality and low shoot growth on all treatments, so only results for callery pear are discussed below. The vertical drain and Enkadrain® treatments produced a significant aeration effect on the nearby soil, whereas the amended backfill and soil trenches did not. This aeration effect did not result in an increase in shoot growth when compared to controls or the other treatments, however. Shoot growth of those treatments designed with a root 'break-out zone' to reduce resistance to root penetration (amended backfill and soil trenches) had significantly more shoot growth than other treatments (vertical drain and control) (Fig. 2). Resistance to root penetration of the surrounding compacted soil was only below the threshold which limits root growth when the soil was very wet (Fig. 3). Only trees which can tolerate prolonged soil wetness, would therefore be able to make significant root growth into the compacted soil.

Root and Shoot Growth Periodicity

Roger Harris, Graduate Student

Shoot growth is easily observed, but the determination of root growth is much more difficult since roots are hidden from view. Knowledge of root growth patterns of trees is important to arborists since it may indicate the best time to transplant trees, fertilize, apply deicing salts, etc. Moreover, if root growth correlates with shoot growth, then periods of high root growth can be predicted by the observance of shoot growth. The purpose of this research was to determine root growth patterns for landscape trees in upstate New York and to determine if root growth can be predicted from observance of shoot growth. In order to determine the relationship between shoot growth and root growth, four species with two distinct types of shoot growth habit were chosen. Green ash (*Fraxinus pennsylvanica*) and scarlet oak (*Quercus coccinea*) grow with more than one flush of shoot growth during the growing season if conditions are favorable. Turkish hazelnut (*Corylus colurna*) and tree lilac (*Syringa reticulata*) commonly grow with only one flush of shoot growth during the growing season. Both groups consist of an easy to transplant tree (green ash or tree lilac) and a difficult to transplant tree (scarlet oak or Turkish hazelnut). This combination of plant material resulted in four possible root growth patterns.

Root growth was determined using two separate devices. The first device was a rhizotron. Two 1 M wide X 1 M deep by 8 M long ditches were excavated, and the sides were fitted with clear 0.6 M wide Lexan plates. The Lexan plates were held by pressure

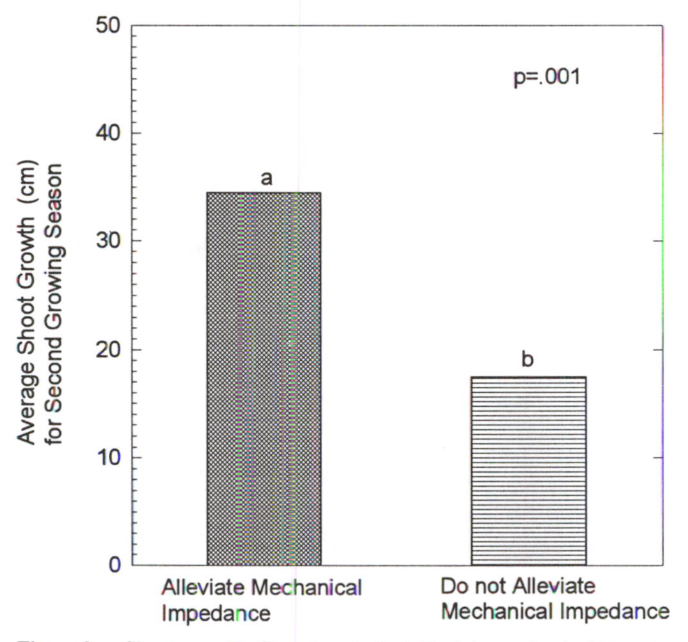

Figure 2 — Shoot growth of treatments that alleviate mechanical impedance (amended backfill and soil trenches) and treatments that do not alleviate mechanical impedance (vertical drain and control). Enkadrain® was considered intermediate and was therefore not considered.

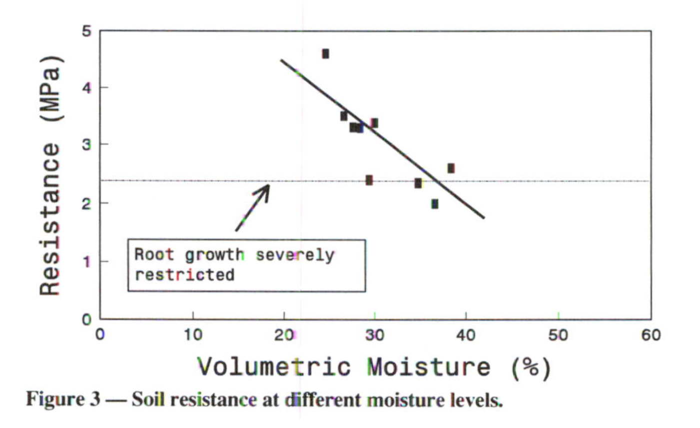

Figure 3 — Soil resistance at different moisture levels.

treated wooden frames, and the frames were supported by a series of posts and beams. The rhizotron was then fitted with plywood covers. Foam insulation was placed against the Lexan plates during the winter. Trees were planted along the edge of the rhizotron and

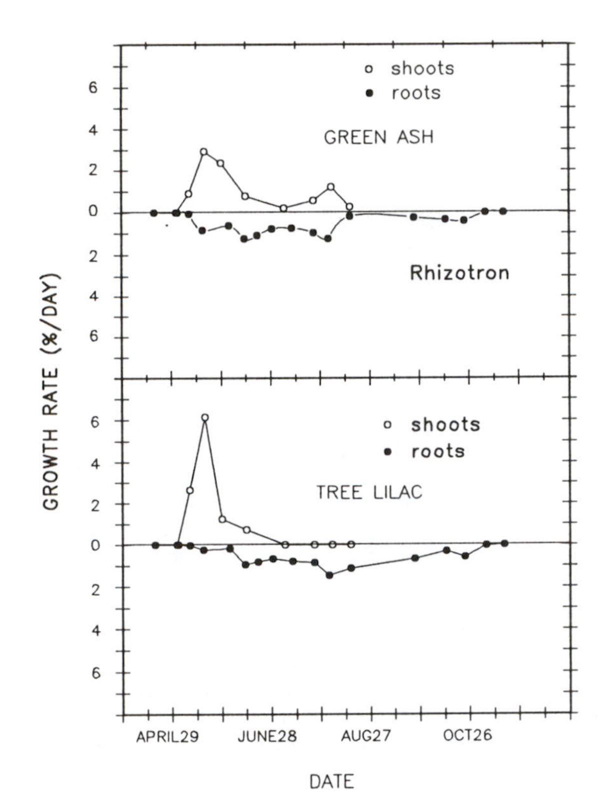

Figure 4 — Shoot and root growth of green ash and tree lilac measured on a rhizotron. Each data point is the mean of 4 trees.

allowed to grow for one year before measurements were made. Root growth was then determined by following the growth of larger individual roots over the growing season.

The second device was a minirhizotron. The minirhizotron consisted of one clear 5 cm diameter acrylic tube per tree placed 0.6 M into the ground at a 4° angle. The tubes entered the ground 0.6 M from the trunks of the trees and sloped toward the tree under the ground. Root growth was determined by slowly lowering a miniature movie camera down the tube and recording the roots present on a video tape.

Ease of transplanting had no relationship to root growth patterns when root growth was assessed by either method, and root growth had no relationship to shoot growth when root growth was determined as described above on the rhizotron (Fig. 4). Alternating root and shoot growth was evident, however, when root growth was determined using the minirhizotron (Fig. 5). This seeming contradiction was because of the different types of roots measured with each method. The rhizotron method measured only the laterally growing, larger, more rapidly expanding roots. The minirhizotron, however, measured all diameters of laterally and vertically growing roots. This indicates that root growth can be predicted by shoot growth when all types of roots are considered. The smaller roots which were included in the measurements made on the minirhizotrons are likely the ones most important for water or mineral absorption. It was interesting that measurable root growth did not begin before budbreak on any of the four species tested.

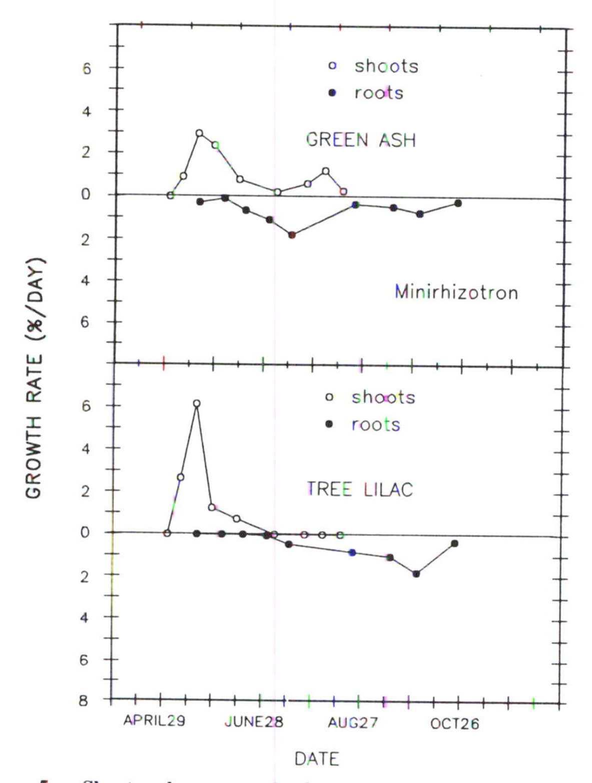

Figure 5 — Shoot and root growth of green ash and tree lilac measured on a minirhizotron. Each data point is the mean of 4 trees.

This was unexpected and contrary to popular thought. This was probably a function of the cold upstate New York soils and the species tested.

Literature Cited

1. Headly, D. B. and N. Bassuk. 1991. Effect of time and application of sodium chloride in the dormant season on selected tree seedlings. J. Environ. Hort. 9(3):130-136.
2. Jaenson, R., N. Bassuk, S. Schwager, and D. Headly. 1992. A statistical method for the accurate and rapid sampling of urban street tree populations. J. Arboric. 18(4):171-183
3. Lindsey, P. and N. Bassuk. 1991. Specifying soil volumes to meet the water needs of mature urban street trees and trees in containers. J. Arboric. 17(6):141-149.
4. Maynard, B. K. and N. L. Bassuk. 1987. Stockplant etiolation and blanching of woody plants prior to cutting propagation. J. Amer. Soc. Hort. Sci. 112(2):273-276.
5. Ranney, T. G., T. H. Whitlow and N. L. Bassuk. 1990. Response of five temperate deciduous tree species to water stress. Tree Physiol. 6:439-438.
6. Sun, W. and N. L. Bassuk. 1991. Stem banding enhances rooting and subsequent growth of M.9 and M. 106 apple rootstock cuttings. HortScience 26(11):1368-1370.
7. Whitlow, T. H., N. L. Bassuk and D. L. Reichert. 1992. A 3-year study of water relations of urban street trees. J. Appl. Ecol. 29:436-450.

Amsterdam Tree Soil

Els A.M. Couenberg

Amsterdam Tree Soil is an artificial soil mixture. It is developed to expand the rooting space of a tree underneath pavements. Amsterdam Tree Soil can be used under pavements that normally bear light loads, especially sidewalks. About 15 years of regular use have proved it to improve tree growth and make good tree growth possible on difficult sites. As an example, applications at the Transvaalkade and the Plantage Middenlaan are shown.

As every urban forester and every landscape architect working in inner cities knows, trees in inner cities don't thrive well. They have an average life span of about 7-10 years, and during that life span often struggle to survive. This is mostly due to a heavily compacted rooting environment of poor quality, in which minimal root growth is possible (1,6,7).

At the end of the sixties, our group, then part of the municipal plantations in Amsterdam, the Netherlands, became aware of the fact that some trees were in poor condition because of unfavorable rooting environment. The first thought, however, was that the trees were suffering from malnutrition. The natural soil at the location where Amsterdam is built was bog-peat. Since bog-peat has no bearing capacity, the upper 2 meters of it have been replaced with medium coarse sand. The city is built on this sand layer. This medium coarse sand has almost no nutritional value.

To find out whether malnutrition was really the cause of the diminished tree growth, a field test was performed on *Tilia*. The sand surrounding the tree was partly replaced with bog-peaty soil, having an organic matter content of about 17%. This was only compacted by foot tamping; the pavement on top of it was put back into place. One year afterwards, the *Tilia* had dark green, fully developed leaves, and a shoot growth of about 40 cm. However, the pavement had settled at least 3 inches (no exact measures were taken of the settlement).

Because the differences in tree growth were well documented and road managers in Amsterdam like trees, this field test resulted in a formal experiment. The road manager, the soil mechanical research department of Amsterdam, our group and the Agricultural University of Wageningen participated. At a test field, pits of 1.2 x 1.2 x 1.2 m were dug out in medium coarse sand in which sand mixed with 2.5% or 5.0% organic matter, respectively, was placed (% by weight, not by volume) and compacted to 2 MPascal. As a control, the medium coarse sand was left in place. Tree growth, soil settlement and several other parameters were measured. The mixture of medium coarse sand containing 5% organic matter (w/w) resulted in the best tree growth, while having acceptable settle-

Els A.M. Couenberg is with OMEGAM Groenadvies, Postbus 94685, 1090 GR Amsterdam, The Netherlands.

ment (19 mm in 3 years compared to the surrounding pavement)(1). This mixture, defined in 1979, and refined after later experiments has since become known as Amsterdam Tree Soil.

In this paper the composition of Amsterdam Tree Soil and the standard tree pit design in which it is used are discussed, followed by some examples of applications.

Composition of Amsterdam Tree Soil

Amsterdam Tree Soil consists of medium coarse sand containing 4-5% (w/w) organic matter and between 2 and 4% (w/w) clay (Fig. 1). It is being made by mixing soils rich in well decomposed organic matter and low in clay content (e.g. former topsoils of bog-peaty pastures) with medium coarse sand using an industrial mixing device (Fig. 2). The medium coarse sand must have a median M_{50} of 220 µm or higher. The uniformity of the sand, called the D60/D10 number, must be lower than 2.5 (3). It must be free of NaCl. It must contain less than 2% particles below 2 µm. These parameters are known to the road engineer and can be given by any soil physics lab.

The organic matter used must be well decomposed to prevent large oxygen consumption due to decomposition of organic matter in the soil mix. The amount of particles below 2 µm must be considerably less than the amount of organic matter in the organic soil used for mixing, estimated by weight. This soil should also be free of NaCl.

The controls of the industrial mixing device can be fine-tuned. The organic matter content of this mix is checked regularly during mixing, to make sure that it keeps within 4-5%. No more than 5% organic matter (w/w) is allowed, to prevent excessive settling of the soil once installed. Having less than 4% organic matter will result in less nutritional value and water for the tree (1,2). Before using a new batch of organic soil, the nutritional value of a test mix is tested. Deficient nutrients are added (almost always potassium and phosphorus) to create a nutrient balance in the initial situation. Nitrogen is not added. After establishing the plant site and planting the trees no nutrients are added.

If more than 4% clay (particles < 2 µm) is used, or if the amount of clay exceeds the amount of organic matter in the mix, it is very easy to compact the soil too heavy during the procedure of filling in and compacting (8).

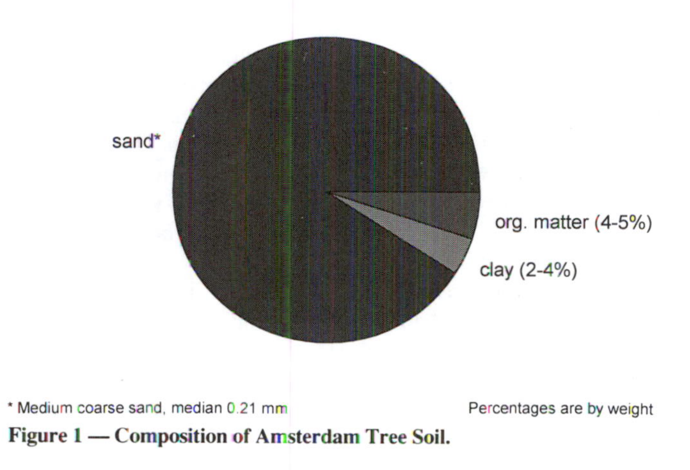

Figure 1 — Composition of Amsterdam Tree Soil.

Figure 2 — Production of Amsterdam Tree Soil using an industrial mixing device. In the right bucket is medium coarse sand, in the left bucket is bog-peaty soil. The piles at the left of the picture are composed of Amsterdam Tree Soil. Nutrients are added by the device connected to the left bucket.

Most of the specifications mentioned above are now part of the RAW Standard Conditions of Contract for Works of Civil Engineering construction (5).

Construction of the Planting Site

Amsterdam has a standard basic planting site design (Fig. 3). It usually has a ground water table of 1 - 1.2 m below ground level. On top of this lies a saturation zone of 10 - 20 cm. On top of that lies a layer of compacted, non-saturated sand. On top of that Amsterdam Tree Soil is filled in and compacted in 2 layers of about 40 - 50 cm, depending of the thickness of the total layer of Amsterdam Tree Soil. This is covered with a compacted layer of 10 cm medium coarse sand for paving. Finally, the pavement is put into place. A typical sidewalk consists of concrete pavers of 30 x 30 x 5 cm. For parking areas, usually bricks of 10 x 20 x 10 cm are used. All medium coarse sand surrounding Amsterdam Tree Soil is compacted to 95-100% Proctor Density.

Amsterdam Tree Soil is *not* compacted to 100% Proctor Density. It is compacted until the soil has a penetration resistance between 1.5 and 2 MPascal (187 to 250 PSI). This is checked during compaction with a penetrometer. It is filled in and compacted in 2 layers, to assure an evenly compaction rate (1,9). Comparison of soil density values after filling with soil density at 100% Proctor Density has shown that soil density of Amsterdam Tree Soil after filling amounts to 70-80% Proctor Density (unpublished data).

Even in situations where a deep ground water table is present, Amsterdam Tree Soil is filled in to a maximum depth of 1 meter. Below one meter depth normally the amount of oxygen influx is too low to make good root growth possible (2,4).

The optimum size of the tree pit filled with Amsterdam Tree Soil is 5 - 20 m^3 depending on species and location. The standard depth of the tree pit is 1 meter. If, for instance, because of a high ground water table, it is not possible to use these measures,

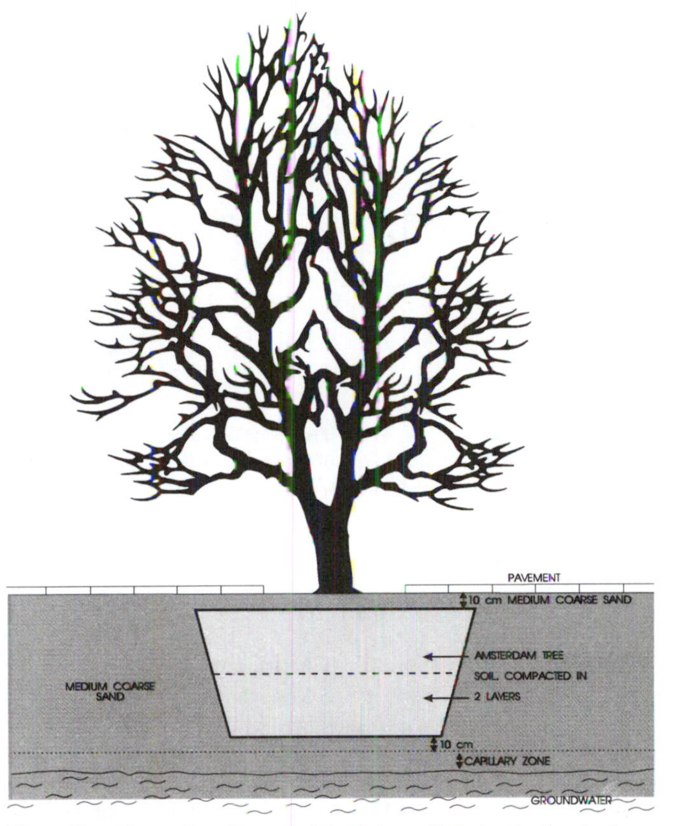

Figure 3 — Current underground basic tree pit design in Amsterdam. The medium coarse sand surrounding the Amsterdam Tree Soil is compacted to >90% Proctor Density. Note that the tree pit is extended underneath the pavement.

the design is adapted so that still as much tree soil as possible can be used. If possible, a trench of about 2 meters wide is used as a shared planting pit along streets (4,9).

Transvaalkade

If a high water table is present, or a standard tree pit is otherwise not possible, other solutions have to be used. In the case of the Transvaalkade (a street alongside a canal in Amsterdam) the maximum possible width of the tree pit was 1 meter. In this case it was decided to grow the trees in a narrow trench at the side of the street where the parking sites were located. The width of the trench was 1 meter, the depth of the tree soil was 0.65 m (Fig. 4). The average amount of tree soil available for each tree is about 4 m^3. The trench was paved with small pavers, leaving raised planting areas for the trees. In 1984 elm trees were planted. In 1988, 4 years after construction, the trees were doing well and had grown considerably (Fig. 5). In 1993 the trees were still doing reasonably well.

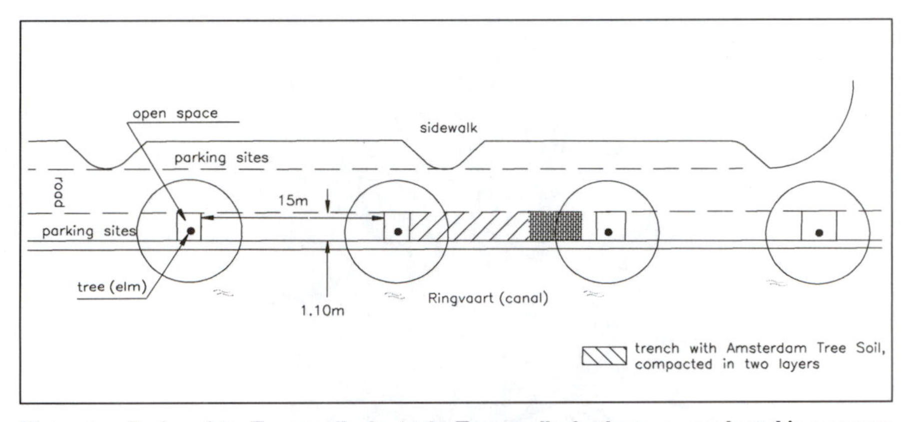

Figure 4 — Design of the Transvaalkade. At the Transvaalkade, the trees are planted in a narrow trench filled with Amsterdam Tree Soil. On top of the trench are parking sites. The pavement on top of the trench consists of small concrete pavers, as is illustrated by the small brick pattern.

Plantage Middenlaan

One of the most outstanding successes of Amsterdam Tree Soil in combination with a clever tree pit design has been the Plantage Middenlaan. This street, entrance to a historical part of the city, containing the Zoo and the botanical garden, had a beautiful elm plantation more than 50 years old. After a sewer renewal, however, they suddenly started to decline (Fig. 6). Experience had taught us that sewer renewal always causes a lot of root damage to adjacent trees. We thus thought that the ratio between roots and shoots was disturbed, and, seeing that the trees were really deteriorating, decided to pollard some of them as a final measure to try to save their lives. This has been done with success in Amsterdam on several other occasions, but in this case it did not help.

Checking the soil conditions revealed that after sewer renewal, the actual rootable soil had diminished to the upper 30 cm. The old sewer was renewed because it leaked. This leaking was a benefit for the trees in that it also functioned as a drain, making rooting possible to a much greater depth. The street was built on a 16th century dike, consisting of heavy clay with an extremely low permeability.

In Amsterdam, being below sea level, all ground water tables are artificial. Proposals to lower the water table of the street were rejected because of possible damage to structures. Due to the original weak bog-peaty subsoil, the houses are built on poles, the older ones on wooden poles. If these wooden poles get above water, they start to rot, causing the valuable houses and other buildings to deteriorate. Thus our ground water table is rigidly controlled and no more than 10 cm deviation is allowed. Fortunately, due to the low permeability of the clay, the planting pits could be drained without damaging structures. The surviving trees, however, could not be saved. They would be replaced with plane trees, to make sure an eventual high water table wouldn't kill the trees immediately and make the high investment worthless.

The new situation was created by digging out a trench 2 meters wide and 1 meter deep in the clay. The bottom of it was filled with gravel in which a double drainage pipe was laid (Fig. 7, a and b). On top of it the pit was filled in with Amsterdam Tree Soil,

Figure 5 — The trees at the Transvaalkade in 1988, 4 years after construction. The trench with the Amsterdam Tree Soil lies underneath the parking sites next to the water. The roots grow underneath the pavement.

Figures 6 — Elms dying due to a high ground water table in the Plantage Middenlaan, summer 1982.

compacted to standard. The drains would be permanently under water, to prevent roots growing into the drains. On top of the Amsterdam Tree Soil a layer of 10 cm compacted paving sand was laid. The whole area was paved either with small pavers, on the parking sites, or special 30 x 30 cm pavers, with wider joints to improve the oxygen influx into the Amsterdam Tree Soil.

Because of the high cost, this was first tried at one end of the street with 8 trees. The drainage system was monitored regularly. With one exception, the system worked perfectly. The trees were growing well. In 1986 the rest of the street was completed. None of the trees has yet had to be removed. They have grown very well and are still growing

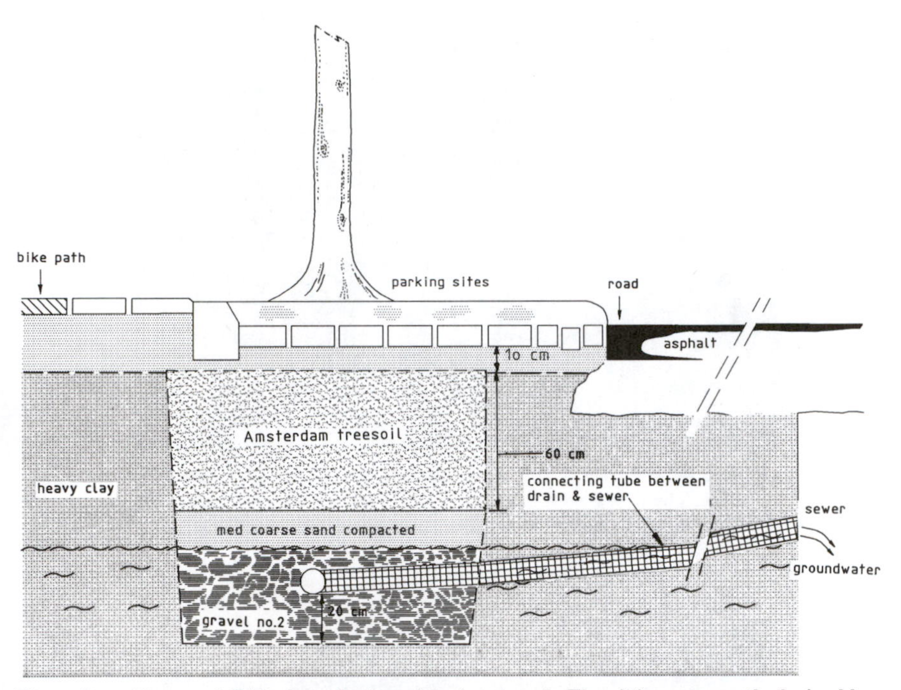

Figure 7a — Plantage Middenlaan: Design of the tree trench. The pit is permanently drained by a drain, lying below groundwater level, so that no roots will grow in it.

(Fig. 8, 9). The drainage system has to be monitored regularly. Once it stops working, which it did once early in the test area, the water level starts to rise quite quickly, thus drowning the trees.

Improving the Soil Around Existing Trees

In cases where older, valuable trees were suspected to have insufficient nutrition, the existing sand has been replaced with Amsterdam Tree Soil. This is expensive, because it can only be done by hand. If the existing soil is removed by a machine, extensive root damage will occur. The normal procedure in these cases is to take out the sand by hand with a shovel and to put in the Amsterdam Tree Soil the same way. To prevent excessive settling, however, the soil has to be compacted. This is done only on top. Fortunately, the existing roots will give some stability and prevent settling to some extent. Experience has shown that the pavement hasn't settled more than expected.

Conclusions and Discussion

The experience with Amsterdam Tree Soil has shown that Amsterdam Tree Soil can be used to expand the tree pit underneath pavements that bear light loads, such as sidewalks. In this way a larger rooting area for urban trees will be available, which will

Figure 7b — Plantage Middenlaan: Construction of the tree trench. To improve percolation, the drain is surrounded by gravel. On top of the gravel Amsterdam Tree Soil is placed. See also figure 7a.

increase the life span and condition of urban trees, especially in inner cities. The experience with Amsterdam Tree Soil has also shown that it is possible to have good tree growth on difficult sites.

However, Amsterdam Tree Soil is no 'wonder mix'. It can be heavily compacted by driving upon it during filling in, especially with heavy machinery. If the tree pit is full of water, the trees will suffocate just as much as when no Amsterdam Tree Soil is present. When there is a drought and no water is given, the trees might die later, but die they will. This has happened in several cases in the Netherlands, where subsoil conditions were not known previous to filling in the soil.

Another point that is not yet known is how long the trees can exist with a given amount of Amsterdam Tree Soil. The Transvaalkade is a good example of this. According to several authors (2,7), medium size trees need at least 5 m^3 soil to provide good growth. From these calculations it can be expected that the growth of those trees will decline any moment. For the next couple of years the tree growth will be monitored to gain insight in how trees will do once maximum expansion is reached in Amsterdam

Figure 8 — Plantage Middenlaan. The trees in 1983, the first summer after finishing the construction. As can be seen, the trench is covered with small pervious concrete pavers.

Figure 9 — Plantage Middenlaan. The back row of the trees of Figure 8 in September 1993, 10 years after construction. The branches in the upper left corner are from the trees of the front row of Figure 8.

Tree Soil. Only time will tell us how long the trees finally will live.

Also, Amsterdam Tree Soil has been used in an area where there is sufficient rainfall. If this soil is used in other climatic areas, adaptations to the tree pit will have to be made after the local situation has been monitored (see also (7)). However, I am convinced that solutions and mixes similar to the ones mentioned above can be applied everywhere in the world. It will take time and effort to get utility and road engineers to accept special solutions and provide adequate space for trees, but the result will be worth fighting for.

Literature Cited

1. Bakker, J.W., C.T. Visser and J.P. Couenberg, 1979. Grondverbetering voor straatbomen. Nota Instituut voor Cultuurtechniek en Waterhuishouding no 1109, 67 pp.
2. Bakker, J.W. 1983. Groeiplaats en watervoorziening van straatbomen. Groen, 1983 no 6, p 205-207.
3. Couenberg, E.A.M. 1993. Bomengrond en de invloed van de zandkorrel. In: Symposium Boom en en Grond, 24 september 1993.
4. Couenberg, J.P., 1983. Zuurstof voor straatbomen; een noodzakelijk gegeven. Groen, 1983 no 6, p 208-209.
5. CROW 1990. Standaard RAW bepalingen (RAW Standard Conditions of Contract for Works of Civil Engineering construction 1990). Paragraph 51.02.03: Grondverbeteren ten behoeve van bomen, p. 503. CROW, Ede, The Netherlands.
6. Kopinga, J. 1991. The effect of restricted volumes of soil on the growth and development of street trees. J. Arboric. 17(3):57-63
7. Lindsey, P. and N. Bassuk 1992. Redesigning the urban forest from the ground below: a new approach to specifying adequate soil volumes for street trees. Arboric. J. 16, 25-39.
8. Luijten, C.J.L.M. and C.J. Mandersloot, 1987. Gedrag van verschillende gronden bij verdichting en de invloed van klei en organisch materiaal hierop. Fysische geografie, RU Utrecht, 49 pp. + appendix.
9. Ros, E.J., 1989. Amsterdamse bomen varen wel bij speciaal grondmengsel. Stad en Groen 3, November 1989.

Transplanting Experiments: What Worked and What Did Not

Michael A. Arnold

The purpose of this article is to bring together the important findings from several studies on root regeneration and transplant establishment in which the author was involved within four states (Ohio, North Carolina, Tennessee, and Texas) over the last seven years. Some techniques have been successful (ex. controlling circling roots with copper treated containers), while others have not (ex. foliar applications of paclobutrazol to delay budbreak and enhance root regeneration), but the attempts always suggested new areas of investigation.

Of the multitude of manipulations necessary to produce, establish, and maintain trees for the urban landscape, the one process common to every tree, except those preserved in situ, is that of transplanting. Bare-root and container-grown trees represent the ends of the spectrum regarding types of planting stock. Bare-root trees are usually in a dormant or quiescent state, lack an intact rootball, and have few if any intact root tips at transplanting. The key to successful transplanting of bare-root stock is the establishment of adequate new roots to support nutrient and water requirements prior to budbreak and resumption of shoot growth. The time frame for this process can range from a few days in late spring to several months in the case of fall planted stock. For container-grown trees the problem is establishment of roots outside of the original container into the surrounding backfill and eventually into the native soils. While container-grown trees usually have an intact rootball and numerous intact root tips, they are also frequently planted during seasons when active leaves are present, thus the demand for nutrient and water uptake are more immediate. Water reserves in the rootball of container materials can be depleted in a single day or less (20). Thus, the overwhelming factor in successful establishment of both types of planting stock is the rapid regeneration of roots.

The seasonality of root regeneration potential further complicates the transplanting process. Root regeneration following transplanting consists of two distinct processes; the adventitious initiation of new roots and the elongation of intact root tips (25). Typically for deciduous temperate zone trees, the capacity for elongation of existing root tips is dependent upon environmental conditions conducive to growth. Thus the potential for root elongation is generally greatest during late spring and fall when soils are warm and moist, may decrease somewhat during some periods of summer due to drought or supraoptimal soil temperatures, and may cease entirely in winter due to cold and

Michael A. Arnold is Assistant Professor, Department of Horticultural Sciences, Texas A & M University, College Station, TX 77843-2133.

unavailability of water in frozen soils. The capacity for root initiation appears to be more closely tied to internal factors in the plant with the periods of greatest potential for root initiation occurring in spring, declining to low levels in the summer, rising again in the fall, and returning to low levels during the cold winter months when trees are dormant. This seasonality of root regeneration is one of the reasons that bare-root stock transplants well in spring and fall when the capacity for root initiation is high and why container-grown stock with their intact root tips can be readily transplanted during the summer months.

This paper will discuss some studies on transplanting, root regeneration and initial plant establishment. The bare-root studies deal mostly with the late winter to early spring time periods corresponding to the rapid seasonal rise in root regeneration potential (Fig. 1) while those studies involving container-grown materials deal with summer and fall transplanting when root regeneration potentials are somewhat lower (Fig. 1). The studies are an amalgamation of experiments performed in the varied environmental conditions of four states, Ohio, North Carolina, Tennessee, and Texas.

Materials and Methods

Bare-root Studies

Root Pruning and Chilling Requirements. This series of experiments (10,11,13) investigated the interactions of winter chilling and root pruning practices on root regeneration potential and the resumption of shoot growth. Dormant fall lifted one-year-old apple liners (*Malus* sp.) with requirements for 1200 to 1600 chilling hours were stored in 5° C cold rooms. Seedlings were planted in containers in a greenhouse (Raleigh, NC) following exposure to increased hours of chilling and various levels of root pruning. New root and shoot growth were monitored during the initial establishment period.

Root Regeneration, Budbreak and Paclobutrazol. Paclobutrazol (1-(4-chlorophenyl)-4, 4-dimethyl-2-(1,2,3-triazol-l-yl)-pentanol) has been shown to reduce internode elongation of several species (21,23,28) and to increase root growth in *Malus x domestica* (apple) and *Vitis vinifera* (grape) (23,24). This experiment was designed to

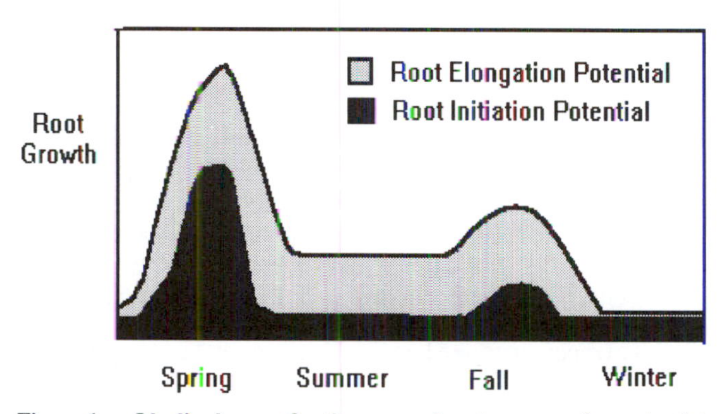

Figure 1 — Idealized curve for the seasonal root regeneration potential of a temperate zone deciduous tree. The lower region represents the potential for root initiation. The upper region represents the potential for root elongation.

test the potential for delay of budbreak and/or increasing root regeneration via the foliar application of paclobutrazol to *Castanea mollissima* (Chinese chestnut) liners. Paclobutrazol was sprayed at 0, 400, 800, or 1200 mg/liter (ppm) to run-off on leaves and stems of two-year-old Chinese chestnut seedlings on August 27 or September 27, 1991, in a sandy loam field plot. Seedlings were fall lifted and placed in 4° C storage until the following spring. On May 5, 1992, the same paclobutrazol treatments were applied to the stems of a previously non-treated group of the stored seedlings. Then, 15 seedlings of each treatment (12 combinations) were planted in a field plot (Cookeville, TN) in a randomized complete block design. Ten additional seedlings of each treatment combination were transplanted to pots set below ground level for use in determining root regeneration characteristics at budbreak and first leaf expansion (2 cm long). First year growth parameters were measured on the field planted seedlings at the end of the growing season.

Antitranspirant Sprays and Hydrogel Dips. In recent years improved antitranspirant formulations for reducing water loss during storage of planting stock (17) have been developed. Also, hydrogels (moisture retaining polymers) for use as root dips or media amendments to reduce moisture stress at transplanting have become available (19,27). This experiment was designed to test the effects of combinations of antitranspirant sprays and hydrogel root dips on water loss during storage and subsequent post-transplant growth of *Quercus rubra* (red oak) seedlings in containers and field sites. On November 11, 1991, seedlings were lifted bare-root, roots washed, and antitranspirant (1:3, vol:vol, Moisturin-4, Burke Protective Coatings, Washougal, WA : water) sprayed on the roots, shoots, whole plant, or no portions of the plant. Each antitranspirant treatment was further separated into three groups, 12 total treatments. The roots of one group were dipped in a hydrogel suspension (3:1, vol:vol, water : Supersorb*F, Aquatrols, Cherry Hill, NJ) after antitranspirant application and then all seedlings were placed in 4° C storage until the following spring. In May, 1992, seedlings were removed from storage, a second subset was dipped in the hydrogel treatment, while a third subset remained non-treated with hydrogel. Ten seedlings of each treatment combination were transplanted to 2.3 liter (#1) nursery containers and grown in standard nursery conditions for the season. Fifteen seedlings of each treatment combination were transplanted to a non-irrigated field plot (Cookeville, TN). Root growth of container-grown plants and shoot growth of all plants were determined at season's end. Xylem water potentials of terminal stems of an additional five plants per treatment were measured prior to and at the end of cold storage.

Warm Root Zone, K-IBA, and DCPTA. The objectives of these experiments were to determine if combinations of warm root zones, cold shoot zones, and application of root growth promoting substances could be used to induce root initiation prior to transplanting of cold stored *Quercus coccinea* (scarlet oak) seedlings. One-year-old fall lifted scarlet oak seedlings were placed in 4° C storage until April, 1992, when the post-storage treatments were applied. Seedlings were placed in 4° C shoot zones with roots in either 4° C or 21° C for 0, 7, 14, or 21 days prior to transplanting. Subsets of the seedlings received either 15 second root dips in 0, 1, 3, or 5 mg K-IBA (potasium salts of indolebutyric acid)/liter of water or 30 minute root soaks in 0, 1, 10, or 100 ppm DCPTA (2-(3,4-dichlorophenoxy)triethylamine prior to temperature treatments. After post-storage treatments, all seedlings were transplanted to 2.3 liter (#1) nursery pots and grown outdoors in standard container nursery conditions (Cookeville, TN). Root regeneration was assessed after completion of the post-storage treatments, at budbreak , and at the end of the first growing season.

Container-grown Trees

Copper Treated Containers and Root Circling. In these experiments (1,2,8,9,12) various species of deciduous temperate zone landscape trees were grown in 2.3 liter (#1) to 46 liter (#5) nursery containers either painted on interior surfaces with various concentrations (0 to 500 g/liter) and formulations ($CuCO_3$, $CuSO_4$, $Cu(OH)_2$) of copper in latex carriers (interior paint, exterior pant, Wiltproof, and Spin Out) or not. Plants were grown under standard nursery conditions. Control of root circling was accessed at from 28 days to an entire growing season from potting.

Mode of Action of Copper Treated Containers. *Fraxinus pennsylvanica* (green ash) seedlings were grown in 2.3 liter (#1) plastic containers painted on interior surfaces with 100 g $CuCO_3$/liter of interior white acrylic latex paint or not treated (6). After 85 days in a greenhouse, the seedlings were harvested and divided into tissue groups for mineral nutrient analysis. Tissues assayed included foliage, stems, main or tap roots, basipetal portion of lateral roots (brown and suberized tissue), mid-section or transition section of lateral roots (mostly succulent with some brown suberized areas), and root tips (succulent and white).

Studies of the effects of alternative formulations of copper (cupric sulfate, cupric carbonate, cupric chloride, and cupric acetate) in woody plant media on the in vitro rooting of *Betula pubescens x papyrifera* microcuttings were conducted to develop dose response curves for 1 to 157 M Cu (<1 to 10 ppm Cu) for root growth and copper toxicity symptoms (5).

Copper Treated Containers, Root Morphology, and Root Regeneration. These experiments were designed to determine the effects of copper treated containers on root morphology and root regeneration potential. In one experiment *Fraxinus pennsylvanica* seedlings were grown inside $CuCO_3$ treated or non-treated container halves affixed to the glass sides of root observation boxes (7). Root elongation rates prior to and after contacting container surfaces, and following release from contact with container surfaces, were monitored. Root system fibrosity or branching was measured by recording distances from root tips to secondary lateral roots and the number of secondary roots formed.

In other experiments (2, 8, 12), green ash, red oak, and apple seedlings were grown in $CuCO_3$ or $Cu(OH)_2$ (100 g/liter) treated or non-treated 2.3 liter (#1) black plastic nursery pots (Columbus, OH; Raleigh, NC; Cookeville, TN) for several months and then transplanted to larger 11 to 46 liter (#3 to #5) treated or non-treated containers. Root regeneration outside the original rootball, root growth inside the original rootball, and shoot growth following transplanting were measured.

Effects of $Cu(OH)_2$ on root distribution were determined by growing nine species of coarse-rooted trees in treated (100 g/liter) or non-treated 2.2 and 11 liter (#1 and #3) nursery pots (2). Rootballs were sectioned into interior versus exterior and upper versus lower halves. Root weights and number of primary lateral roots in each section were recorded at the middle and end of the growing season (Cookeville, TN).

Copper Treated Containers and Transplant Establishment. Several tree (green ash, red oak, and red maple (*Acer rubrum*)) and herbaceous bedding plants (*Impatiens wallerana* (impatiens), *Celosia cristata* (celosia), *Pelargonium x domesticum* (geranium), *Tagetes erecta* (marigold), *Chrysanthemum x superbum* (shasta daisy), and *Coreopsis lanceolata* (coreopsis)) were grown in $CuCO_3$ or $Cu(OH)_2$ treated (25 to 100 g/liter) or non-treated containers (1,3,6,8). Woody plants were root pruned (four 2.5 cm deep vertical slashes down the sides of the rootball and matted roots at bottom of the container removed) or not to correct circling roots prior to transplanting to containers and field sites. Growth assessments were made at the end of container production, at the end of the growing season for the woody plants, and after 35 days in the field for the herbaceous bedding plants. Photosynthesis and transpiration of *Fraxinus pennsylvanica* and leaf xylem water potentials of *Acer rubrum* seedlings were measured (1,6).

Results and Discussion

Bare-root Studies

Root Pruning and Chilling Requirements. Two factors regarding transplanting of fall lifted bare-root trees held in winter storage significantly impacted initial shoot growth. First, there was an increase in shoot growth as chilling hours increased (Fig. 2) (10,11,13). This is not surprising as it has been long recognized that resumption of vigorous shoot growth was dependent on satisfying cold dormancy chilling requirements (22). When storing apple and peach (*Prunus persica*) species this does not normally pose a problem as their chilling requirements are usually less than 1500 hours at 5° C. However, the chilling requirements of other landscape species such as sugar maple (*Acer saccharum*) can be in the range of 3000 hours (26). Thus, in regions where little chill unit accumulation occurs prior to storage or after spring planting it is important to fully satisfy the chilling requirements during storage. The idea is to transplant plants at the height of the spring increase in root initiation potential (Fig. 1).

The second factor is of greater general concern since our studies indicated that the removal of small lateral roots can greatly decrease shoot growth after transplanting of stored bare-root stock (Fig. 2). During lifting and common storage, desiccation of bare-root planting stock is a common problem (16). Desiccation of small roots could likely result in similar responses as pruning removal of those same roots. Thus, every effort should be made to preserve as many small lateral roots as possible.

Root Regeneration, Budbreak and Paclobutrazol. Late summer applications of paclobutrazol at 1200 mg /liter and fall applications of 800 mg/liter increased the num-

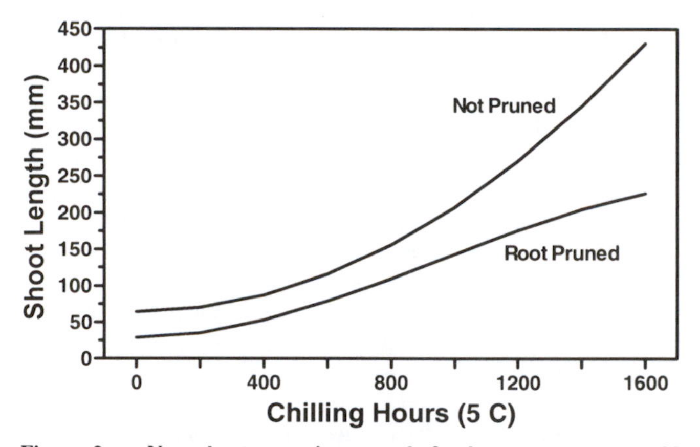

Figure 2 — New shoot extension growth for bare-root one-year-old *Malus x domestica* seedlings transplanted following 0 to 1600 chilling hours at 5°C, either with small lateral roots removed at transplanting or not.

ber of small (< 1 mm diameter) roots regenerated at budbreak and across application times. Four hundred to 800 mg/liter increased the number of large roots regenerated at first leaf expansion. Also, budbreak was delayed by one to three days with 400 to 1200 mg/liter applications made prior to fall lifting. However, none of these responses was of sufficient magnitude to improve growth or survival during the first growing season after transplanting (data not presented). Greater concentrations of paclobutrazol, above 1200 mg/liter, could be tested, but reductions in internode growth during the first growing season were noted on some 1200 mg/liter treatments.

Antitranspirant Sprays and Hydrogel Dips. Results of antitranspirant sprays and hydrogel dips were similar to those of the paclobutrazol studies. A combination of spraying the whole plant with antitranspirant and dipping the roots in the hydrogel prior to storage did reduce water loss during storage (-0.5 MPa stem xylem potential versus -0.8 MPa to -1.5 MPa for the other treatment combinations), the reduced water loss did not result in any growth improvements either in containers or in the field (data not presented). In fact, some hydrogel dipped seedlings grown in containers exhibited what appeared to be a form of root rot, suggesting that in continually moist media, hydrogels may actually be detrimental. Reductions in root and shoot growth of container stock grown in pine bark based media amended with hydrogels has been reported (19,27).

Warm Root Zone, K-IBA, and DCPTA. In previous studies on interactions between root regeneration and dormancy in apple rootstocks (13), it was noted that combinations of K-IBA root dips and temperature conditions of 5° C shoots and 20° C roots following cold storage resulted in high levels of root regeneration while shoots remained in a dormant or quiescent state (Fig. 3). The present study was designed to determine if it would be possible to identify a combination of K-IBA concentration and post-storage warm root/cold shoot combination that would result in large numbers of roots that had initiated, but not yet emerged from within the mother root at the time of transplanting to the field. The hypothesis was that this would result in greater quantities of new roots to support emerging shoot growth. Results with scarlet oak indicated that visible root regeneration increased with increased K-IBA concentration and increased time in warm root zones (Fig. 4A). However, those treatment combinations with numerous visible new roots at transplanting did not have as many regenerated roots at budbreak as the 3 to 5 mg K-IBA/liter and 7 to 14 day warm root zone treatments (Fig. 4B). Despite considerable care during transplanting and the relatively favorable container nursery conditions, many of the initiated roots that had already emerged from the mother roots at transplanting were damaged during transplanting. Under the favorable conditions in a container nursery this increased root regeneration at budbreak had little effect on shoot growth during establishment (data not presented), however under more demanding field conditions this increased root regeneration may have been more important to growth and/or survival, suggesting more research on this subject is needed. DCPTA root soaks resulted in no commercially significant increases in root regeneration or shoot growth (data not presented).

Container-grown Trees

Copper Treated Containers and Root Circling. In general concentrations of about 100 g/liter of either $CuCO_3$ or $Cu(OH)_2$ reduced circled and matted roots at container wall - media interfaces to commercially acceptable levels on a wide range of tree species (1,2,7,8,9,14) and on a limited number of foliage and bedding plants tested thus far (3, 15). With most bedding plants lower concentrations (25 to 50 g $Cu(OH)_2$ /liter) were

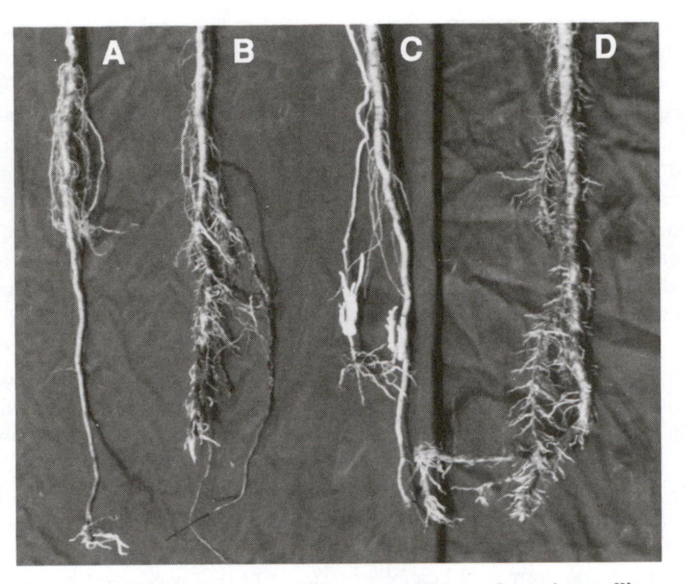

Figure 3 — Roots of one-year-old bare-root *Malus x domestica* seedlings following 1600 hours of chilling at 5°C prior to placement in 20°C shoot and 20°C root zones (for 14 days) after 15 second root dips in 0 (A) or 10,000 (B) mg K-IBA/liter of water (ppm) or placed in 5°C shoot and 20°C root zones following root dips in 0°C(C) or 10,000 (D) mg K-IBA/liter of water.

sufficient to inhibit root elongation (3). Complete control of root circling did not occur with 100 g $Cu(OH)_2$/liter with some very coarse rooted species, such as *Carya illinoinensis* (pecan), *Carya glabra* (pignut hickory), and *Taxodium distichum* (baldcypress), but circled roots were reduced (2,4,9).

Mode of Action of Copper Treated Containers. Elongation of green ash roots was inhibited within three days of contact with $CuCO_3$ treated container surfaces (7). Roots contacting copper treated surfaces (Figure 5A and 5B) exhibited typical symptoms of a mild copper toxicity (swollen, slight brownish to blackish discoloration of the root tips, and reduced elongation) compared to those contacting non-treated surfaces (Fig. 5C). Root tips of birch microcuttings exposed to increasing levels of various copper formulations (5) exhibited similar symptoms (Fig. 5D). Tissue analysis for mineral nutrients revealed a gradient of decreasing copper concentration from root tips to foliage (Table 1). Tissue concentrations of copper were at moderate to mildly toxic levels in the root tips and transition sections of the roots (Table 1). Copper concentrations were reduced to within normal levels within a few cm of the root tips (7). Similar results were reported for *Viburnum* sp. with $Cu(OH)_2$ (18). Thus, evidence suggests that inhibition of root elongation is via a mild copper toxicity confined to the first few cm of the root tips.

Copper Treated Containers, Root Morphology, and Root Regeneration. In root observation box studies, elongation of green ash roots contacting treated surfaces (Figs. 5A and 5B) approached zero within 3 days (7), while those contacting non-treated surfaces continued to elongate (Fig. 5C). Inhibition of root tips by treated surfaces stimu-

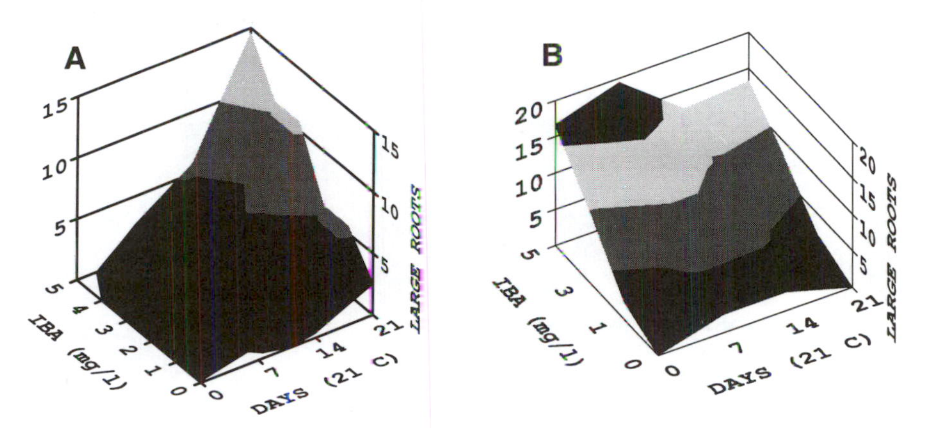

Figure 4 — Effects of increasing days in 4°C shoot and 21°C root zones and increasing concentrations of K-IBA root dips (15 seconds) following cold starage of one-year-old *Quercus coccinea* seedlings on the number of large (>1mm in diameter) roots regenerated during post-storage priming treatments (A) and at budbreak following transplanting (B).

lated branching of subordinate lateral roots (Figs. 5A and 5B). Growing *Quercus falcata* var. *pagodifolia* (cherrybark oak), *Quercus alba* (white oak), *Quercus acutissima* (sawtooth oak), *Juglans nigra* (black walnut), and baldcypress in $Cu(OH)_2$ treated nursery containers resulted in a shift in dry matter allocation from exterior to interior or lower to upper portions of the rootball (9). Root initiation of primary lateral roots (the main structural roots) was greater on the upper half of the tap roots of cherrybark oak, post oak (*Quercus stellata*), and Chinese chestnut seedlings grown in $Cu(OH)_2$ treated versus non-treated containers (9). Seedlings of green ash and apple grown in $CuCO_3$ treated containers had approximately twice the number of white root tips compared to those from non-treated containers (12).

When $CuCO_3$ or $Cu(OH)_2$ was used at 100 g/liter concentrations few root tips were killed, and with green ash most resumed elongation rates comparable to non-treated plants within three to six days of release from contact with copper treated surfaces (7). In species with lower copper tolerances root tips could be killed, but root branching was still increased due to a proliferation of higher order lateral roots behind the tips. Root regeneration of red oak (Fig. 6A) and green ash (Fig. 6B) following transplanting to larger containers was significantly increased over those from non-treated containers (8). The overall effects were that plants from copper-treated containers exhibiting a more fibrous root system that was more evenly distributed within the container volume (Fig. 6C) and less concentrated near external surfaces. When recommended (20) root pruning practices for correcting circled roots was performed substantial portions of the lateral roots were lost with non-treated trees, while those from copper treated containers needed little or no corrective pruning (Fig. 6C) resulting in a more fibrous intact root system at transplanting.

Copper Treated Containers and Transplant Establishment. The final measure of any "improved production practice" is whether the change resulted in better establishment of the plants in the landscape. Increases in growth and/or reductions in visible symptoms of transplant shock (leaf senescence, defoliation, dieback) in the field have been reported for red oak (8), green ash (8), red maple (6), and impatiens (3) for plants

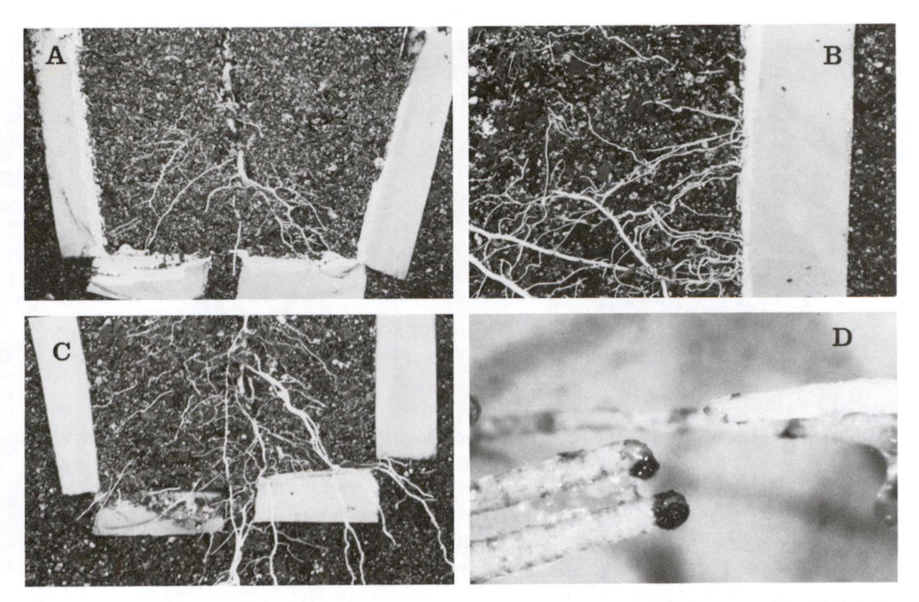

Figure 5 — Root growth of one-year-old *Fraxinus pennsylvanica* seedlings grown for 27 days in root observation boxes with 2.3 liter (#1) nursery container halves affixed to the glass panels after painting the interiors with 100 g $CuCO_3$/liter of white interior acrylic latex paint (A and B) or not (C). Root tips of *Betula pubescens* x *papyrifera* microcuttings grown *in vitro* for 28 days in test tubes containing woody plant media with 1 M Cu (D, root on right) or 118 M Cu (D, roots on left) as $CuSO_4$*$5H_2O$.

Table 1. Copper concentration (mg/liter, ppm) in various tissues of *Fraxinus pennsylvanica* seedlings grown in 2.3 liter (#1) black plastic nursery containers painted on interior surfaces with 100 g $CuCO_3$/liter white interior latex acrylic paint or not

Tissue	Non-treated	Cu-treated
Foliage	2	6
New stems	2	12
Old stems	3	12
Tap roots	4	16
Lateral roots		
Basal section	6	41 *
Transition section	14	169 *
Root tips	27	493 **

* Indicates luxury consumption to mild toxicity concentrations.
** Indicates moderate to very toxic concentrations.

transplanted from copper treated containers compared to those from non-treated containers. Controlling root circling with $CuCO_3$ treated containers resulted in less severe reductions in net photosynthesis (Fig. 7A) and transpiration (Fig. 7B) in green ash trees than correcting circling roots at transplanting by root pruning plants from non-treated containers (1). Diffences in net photosynthesis and transpiration (Figs. 7A and 7B) were

Figure 6 — *Fraxinus pennsylvanica* (A) and *Quercus rubra* (B) seedlings in November that were transplanted in May from 2.3 liter (#1) containers, painted on interior surfaces with 100 g $CuCO_3$/liter of white interior acrylic latex paint (plant on left) or not (plant on right), to non-treated 46 liter (#5) containers. *Fraxinus pennsylvanica* (C) grown in 2.2 liter (#1) containers painted on interior surfaces with 100 g $CuCO_3$/liter of paint (plant on right) or not (plant on left). The center plant in C was grown in a non-treated container, but root pruned according to recommended practices to correct for circling roots.

observed for the critical first few weeks of the establishment period with both summer and early fall transplanting (1). A similar less severe reductions in leaf xylem water potential were observed for trees transplanted from $Cu(OH)_2$ treated containers compared to those from non-treated containers (6).

Summary

What is the bottom line from all these experiments? The take home message with bare-root materials are to preserve as many lateral roots as possible because they are critical to maximizing shoot growth following transplanting. It is also important in warmer regions to ensure that bare-root common storage stock's chilling requirements have been met prior to outplanting. The area of promise for future research may be the use of post-storage priming treatments for bare-root stock.

For container-grown materials the most important findings are the potential to control root circling without compounding transplant shock by root pruning. For some species copper-treated containers also appear to improve initial establishment. While numerous effects of copper-treated containers have been documented, questions remain. Can copper-treated containers be adapted for use with other crops (ex. bedding plants, house plants, vegetable transplants) or to improve the success of other production techniques (ex. production stage inoculation with mycorrhizal fungi)? What is the potential for copper leaching or accumulation in recycled irrigation water in nurseries? Can copper-treated containers be used to extend the shelf-life of various classes of container-grown plants (i.e. nursery stock, bedding plants, house plants)? Can copper containing latex compounds be used to control root growth in landscape situations (ex. copper-treated edging materials)? What are the potentials for integrating provenance or

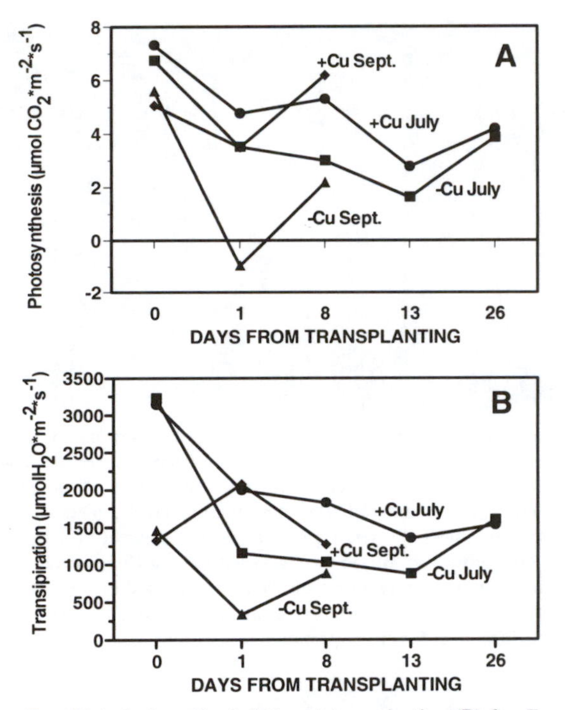

Figure 7 — Net photosynthesis (A) and transpiration (B) for *Fraxinus pennsylvanica* seedlings transplanted on July 15 or September 15 from 2.3 liter (#1) containers painted on interior surfaces with 100 g $CuCO_3$/liter of white interior acrylic latex paint (+) or not (-). Seedlings from non-treated containers were root pruned to correct circling roots at transplanting.

improved seed source selection into the existing production and transplant techniques (preliminary field studies suggest 20 % or more growth increases (4))? While these studies have answered a number of important questions they have raised as many more. Progress begins by asking a question.

Acknowledgment

Technical Article No. TA 31344 of the Texas Agricultural Experiment Station (TAES). The author thanks Larry Shoemaker, Randall Culbertson, Renee Simms, Joseph Takayama, Edgar Davis and Robert Belding for their technical assistance and Dr. Henry Yokoyama for providing the DCPTA samples. The use of trade names in this publication does not imply endorsement by the TAES, Texas A & M University or the author of the products named, nor criticism of similar ones not mentioned.

Literature Cited

1. Arnold, M.A. 1987. Cupric carbonate modification of *Quercus rubra* and *Fraxinus pennsylvanica* root systems and implications for production and transplant. M.S. Thesis, The

Ohio State Univ., Columbus, Ohio, p. 215.

2. Arnold, M.A. 1992. Timing, acclimation period, and cupric hydroxide concentration alter growth responses of the Ohio production system. J. Environ. Hort. 10:114-117.
3. Arnold, M.A., D.L. Airhart, and W.E. Davis. 1993. Cupric hydroxide-treated containers affect growth and flowering of annual and perennial bedding plants. J. Environ. Hort. 11:106-110.
4. Arnold, M.A. and W.E. Davis. 1992. Provenance selection improves the nursery performance of sycamore and sweet gum seedlings in the field. Proc. SNA Res. Conf. 37:137-139.
5. Arnold, M.A., R.D. Lineberger, and D.K. Struve. 1994. Influence of copper compounds on in vitro rooting of birch microcuttings. J. Amer. Soc. Hort. Sci. 119:74-79.
6. Arnold, M.A., G.K. Stearman, and R.W. Cripps. 1992. Post-transplant water relations and growth responses of red sunset red maples trees to root pruning and $Cu(OH)_2$-treated container production. HortScience 27:569(007).
7. Arnold, M.A. and D.K. Struve. 1989. Cupric carbonate controls green ash root morphology and root growth. HortScience 24:262-264.
8. Arnold, M.A. and D.K. Struve. 1989. Growing green ash and red oak in $CuCO_3$-treated containers increases root regeneration and shoot growth following transplant. J. Amer. Soc. Hort. Sci. 114:402-406.
9. Arnold, M.A. and D.K. Struve. 1993. Root distribution and mineral nutrient uptake of coarse-rooted trees grown in cupric hydroxide-treated containers. HortScience 28:988-992.
10. Arnold, M.A. and E. Young. 1990. Differential growth responses of apple species to chilling and root pruning. J. Amer. Soc. Hort. Sci. 115:196-202.
11. Arnold, M.A. and E. Young. 1990. Growth and protein content of apple in response to root and shoot temperature following chilling. HortScience 25:1583-1588.
12. Arnold, M.A. and E. Young. 1991. $CuCO_3$-painted containers and root pruning affect apple and green ash root growth and cytokinin levels. HortScience 26:242-244.
13. Arnold, M.A. and E. Young. 1991. Influence of chilling at 5 C, IBA, suckering and top growth on root regeneration in seedlings of *Malus* spp. J. Hort. Sci. 66:423-433.
14. Beeson, R.C. and R. Newton. 1992. Shoot and root responses of eighteen southeastern woody landscape species grown in cupric hydroxide-treated containers. J. Environ. Hort. 10:214-217.
15. Case, G.N. and M.A. Arnold. 1992. Cupric hydroxide-treated containers decrease pot binding of five species of vigorously rooted greenhouse crops. Proc. SNA Res. Conf. 37:94-98.
16. Davidson, H., R. Mecklenburg, and C. Peterson. 1988. Nursery Management Administration and Culture, 2nd Ed. Prentice Hall, Englewood Cliffs, N.J. p. 413.
17. Englert, J.M., L.H. Fuchigami, and T.H.H. Chen. 1991. Reducing water loss from bare-root nursery trees after harvest. HortScience 26:732(365).
18 Flanagan, P.C. and W.T. Witte. 1991. Effects of chemical root pruning on root regeneration and cellular structure of viburnum root tips. Proc. SNA Res. Conf. 36:46-49.
19. Keever, G.J., G.S. Cobb, J.C. Stephenson, and W.J. Foster. 1989. Effect of hydrophylic polymer amendment on growth of container grown landscape plants. J. Environ. Hort. 7:52-56.
20. Harris, R.W. 1992. Arboriculture, 2nd Ed. Prentice Hall, Englewood Cliffs, N.J. 674.
21. Hunter, D.M. and J.T.A. Proctor. 1992. Paclobutrazol affects growth and fruit composition of potted grapevines. HortScience 27:319-321.
22. Perry, T.O. 1971. Dormancy of trees in winter. Science 171:29-36.
23. Sharma, S. and A.D. Webster. 1992. The effects of growth regulator sprays applied in the

nursery on apple scion growth and the induction of roots on M 9 rootstock stems and tree anchorage. Gartenbauwissenschaft 57(4):173-177.

24. Smith, E.F., I. Gribaudo, A.V.Roberts, and J. Mottley. 1992. Paclobutrazol and reduced humidity improve resistance to wilting of micropropagated grapevine. HortScience 27:111-113.
25. Sutton, R.F. and R.W. Tinus. 1983. Root and root system terminology. For. Sci. Monograph 24. p. 137.
26. Taylor, J.S. and E.B. Dumbroff. 1975. Bud, root, and growth regulator activity in *Acer saccharum* during the dormant season. Can. J. Bot. 53:321-331.
27. Tripepi, R.R., M.W. George, R.K. Dumbroese, and D.L. Wenny. 1991. Birch seedling response to irrigation frequency and a hydrophilic polymer amendment in a container medium. J. Environ. Hort. 9:119-123.
28. Vu, J.C.V. and G. Yelenosky. 1992. Growth and photosynthesis of sweet orange plants treated with paclobutrazol. J. Plant Growth Regulation 11:85-89.

The Effect of Treeshelters on Shoot/Root Development of Some Ornamentals Grown in Containers or in the Landscape

Pavel Svihra, David W. Burger and Richard Harris

This study was designed to test the efficacy of treeshelters on selected container-grown trees in the nursery and on redwood seedlings transplanted into the landscape. Treeshelters accelerated shoot growth of southern magnolia, holly oak and deodar cedar but root growth was reduced during the first growing season as compared to the controls. Sufficient shoot/root development of the trees was achieved only after two growing seasons. One-year-old redwood seedlings successfully established themselves whether grown in treeshelters or not, while receiving 1/7 to 1/14 as much water as they would in a nursery bed grown in one-gallon containers. The seedlings in treeshelters grew significantly taller.

For the $450 million worth of woody ornamental plants sold each year by California nurseries, treeshelters (cylindrical polypropylene tubes about 10 cm, or 4 inches, in diameter and varying heights [Fig.1]) with their inter-environment accelerating tree growth, could save money and water (2,3,4). We hypothesized that the system could have a potential in the California nursery industry that grows most of the woody ornamentals in containers. Additionally, the container environment provided a unique opportunity to study the treeshelter's effect on shoot and root development. We conducted studies to 1) determine how trees would respond to being grown in treeshelters in nursery containers, 2) monitor the environment in and around treeshelters used with container-grown trees, 3) measure the water use of trees in containers grown with or without a treeshelter, and 4) compare the growth and survival of seedlings replanted into the landscape, grown with or without a treeshelter and receiving 7 or 14 times less water than in the nursery.

Trees in Nursery Containers

Methods

Study of treeshelter application in the nursery included *Cedrus deodora*, *Quercus ilex*, and *Magnolia grandiflora* (30 of each) (Fig.1). The nursery industry identified these species as difficult to grow with poor survivorship and growth. We randomly

---

Pavel Svihra is with the University of California, Cooperative Extension, 1682 Novato Blvd. Suite 150 B, Novato, CA 94947. David W. Burger and Richard Harris are with the Department of Environmental Horticulture, University of California, Davis, CA 95616

selected 30 1-gallon containers from a pool of several hundred and then replanted them into 5-gallon containers.

We measured (methods described in Burger et al. [1]): a) plant height and trunk caliper at replanting in February 1990, after the first growing season, and after the second growing season; b) temperature, relative humidity (Fig.2) and CO_2 concentrations inside and outside the shelter (Fig.3); c) water consumption twice during the first grow-

Figure 1 — Partial view of experimental setting in the nursery.

Figure 2 — Temperature and relative humidity inside and outside of the shelter was measured with HI 8564 thermohygrometer; Hanna Instruments, Singapore.

ing season by watering the plants to the full saturation of the containers, then allowed them to drain for one hour. We then weighed each plant with container and again weighed them after 24 hours. The difference was defined as the water used through transpiration by the plant and evaporation from the soil surface. We determined the fresh and dry weights of the tops and roots of randomly selected after the first growing season and all experimental plants after the second growing season.

Results

Fig. 4 diagrammatically shows that treeshelters raised the temperature 3°C (5.5°F) and relative humidity 21% and the CO_2 was significantly greater inside the shelter near the ground. The height of all three species grown in shelters for 1 year was significantly greater than that of unsheltered controls (Table 1). The height increase difference between sheltered and unsheltered *Quercus* trees continued through the second year but did not for *Cedrus* and *Magnolia*.

Cedrus trees grown in shelters used significantly less water than those grown without shelters. Even though *Magnolia* and *Quercus* trees in shelters used less water than those without shelters, the differences were not statistically significant. Dry weights comparing root development in treeshelters were negatively affected in the first growing season (Table 1).

Figure 3 — The needle of a 1-ml syringe inserted through the septum to the inside of the treeshelter helped to collect gas samples for analysis of CO_2 concentration in Horiba PIR-2000 infrared CO_2 analyzer.

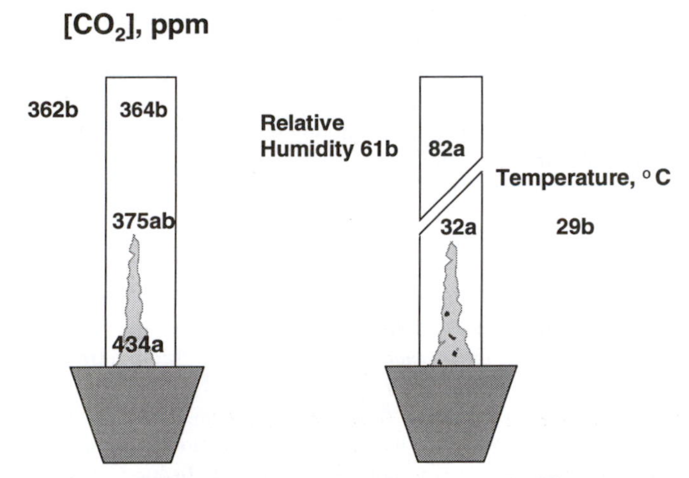

Figure 4 — The effect of treeshelter environment on the CO_2 concentration, relative humidity and temperature.

The effect of treeshelters on shoot and root growth was not significantly different after the second growing season, and we compared control vs staked, control vs in the shelter for two years, control vs in the shelter for one year and then staked, and control vs in the shelter for one year and then unstaked.

Quercus trees staked and unsheltered for two years were shorter than those that had grown in shelters for two years. Staked *Quercus* trees had a smaller caliper diameter than unstaked trees. Neither fresh weights nor dry weights of shoots and roots was significantly affected by treeshelter treatments after two years (Table 1).

Landscape Trees in a Park

Application of treeshelters for container-grown woody ornamentals stimulated new research in the landscape.

Method

Fifteen seedlings of *Sequoia sempervirens* were planted in the landscape. Seedling height and trunk caliper were measured at planting time, on April 10, 1991 and on May 27, 1992. Ten treeshelters were placed over randomly selected seedlings; five marked with a dot and five with an asterisk. Seedlings were checked weekly during the growing season and biweekly in fall and winter.

During the establishment growing season the seedlings in the experiment were subjected to the irrigation schedule shown in Fig.5. This schedule remained until the first rain occurred on October 29, 1990. After that, the plants were left to adjust their growth and survival to the water supply from natural precipitation. In November 1993 shoot fresh/dry weights and root fresh/dry weights between each treatment will be compared. The root architecture for all three treatments will also be evaluated.

Results

Redwood seedlings protected by treeshelters and irrigated with 1 liter of water every

Table 1. Growth of three tree species after two years grown with and without treeshelters for 1 or 2 years. Values in each parameter category for each species followed by the same letter are not significantly different at p=0.05 using Scheffe's Mean Separation Test. SFW—Shoot Fresh Weight, SDW—Shoot Dry Weight, RFW—Root Fresh Weight, RDW—Root Dry Weight, NS—Not Significant*

Treatment	Height (cm)	Caliper (mm)	SFW (g)	SDW (g)	RFW (g)	RDW (g)
			Cedrus			
No stake, no shelter	170 B	34.7	2710	1314	2357	888
No stake, shelter - 1 Year	211 A	35.3	2623	1264	2271	862
No stake, shelter - 2 Years	212 A	29.7	2293	1092	2373	809
		NS	NS	NS	NS	NS
			Quercus			
No stake, no shelter	183 C	29.0 A	1600	966	1425	669
No stake, shelter - 1 Year	242 B	30.0 A	1906	1175	1295	614
No stake, shelter - 2 years	271 A	30.0 A	1763	1040	1266	603
Staked, no shelter	221 BC	18.5 B	-	-	-	-
			NS	NS	NS	NS
			Magnolia			
No stake, no shelter	116 B	19.0	1170	536	659	223
No stake, shelter - 1 Year	163 A	15.3	640	325	434	148
No stake, shelter - 2 Years	176 A	19	1043	487	565	203
		NS	NS	NS	NS	NS

*(From Svihra, P. et al. 1993, by permission of *California Agriculture*)

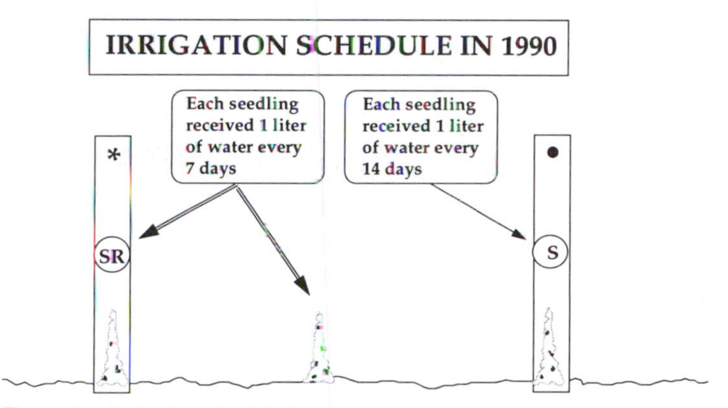

Figure 5 — Irrigation schedule for *Sequoia sempervirens* at the establishment growing season in the landscape.

14 days (S-schedule) grew significantly higher than unprotected control seedlings, which received 1 liter of water every 7 days. Seedlings with SR-schedule were about 60 to 63% taller than the unprotected control seedlings under the same watering schedule, but this difference was not significant (Fig. 6).

Neither treeshelters nor watering schedule had a significant effect on caliper,

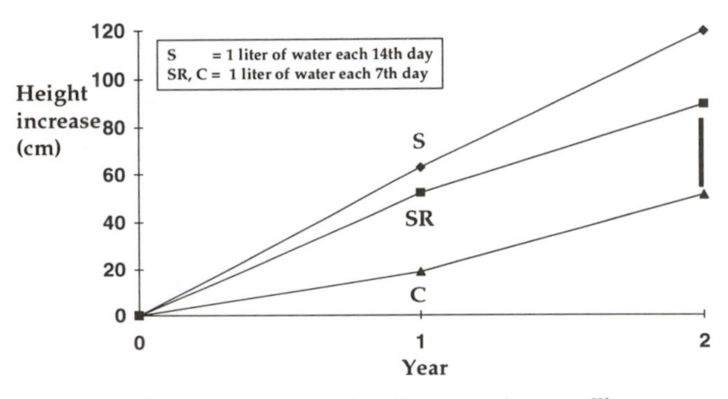

Figure 6 — Growth response of *Sequoia sempervirens* seedlings grown with and without treeshelters after each seedling in group "S" received 1 liter of water each 14th day; group "SR" received 1 liter of water each 7th day; and group "C" control unprotected seedlings received a similar schedule as the "SR" group. The vertical bar indicates one standard deviation.

Figure 7 — The effect of treeshelter on the root development of hydroponically grown *Populus nigra*.

although in the second growing season, control seedlings had about 26 to 31% larger caliper than those in the treeshelters.

Fig. 7 shows the difference in root development of *Populus nigra* grown hydroponically in the shelter and unprotected. We are in the process of uncovering the root architecture of redwoods to find out whether this phenomenon affects root development after the plants are grown out of the treeshelters.

Literature Cited

1. Burger, D.W., P. Svihra,and R. Harris. 1992. Treeshelter use in producing container grown trees. HortScience 27(1):30-32.
2. Evans, J. and M.J. Potter. 1985. Treeshelters—a new aid to tree establishment. Plasticulture 68:7-20.
3. Frearson, K. and H.D. Weiss. 1987. Improved growth rates with treeshelters. Quart. J. For. 81(3):184-187.
4. Potter, M.J. 1988. Treeshelters improve survival and increase early growth rates. J. For. 86(6):39-41.
5. Svihra, P., D.W. Burger, and R. Harris. 1993. Treeshelters for nursery plants may increase growth, be cost effective. California Agriculture 47(4):13-16.

Root Development After Transplanting

Gary W. Watson

Root Loss During Transplanting

Root Development in the Nursery

Root growth is subject to the physical and chemical properties of the soil environment. Vertical and horizontal root penetration is dependent upon soil texture, structure and profile. Soil moisture and aeration, which are inversely related, also have a major impact on root growth (26,39). Root and top growth may not occur simultaneously.

The highest root densities are usually found near the soil surface (54). Most roots of nursery-grown trees are found in the top 30 cm of soil (21,22,66). Competition with turf and other vegetation can reduce density very near the soil surface (65,76). For nursery grown trees, roots 1.0 cm in diameter and larger were most developed at the soil depth of 13-38 cm (Figure 1), and with greatest root development in the north quadrant (Figure 2). The spread of the root system is typically up to three times the spread of the branches (16,66). The small fibrous roots which absorb water and nutrients are highly concentrated in the fertile topsoil and occasionally in bands of richer soil at greater depths (52).

Root Loss During Harvesting

Given the spreading nature of the root system, it should not be surprising that only a small fraction of the root system is contained in the root ball when the tree is dug from the nursery. It has been estimated that only 3.8-8.5 percent of the root system (Figure 3) (17,71) and two percent of the soil volume occupied by the root system (66) in the nursery is contained in the root ball. Root pruning inside the root ball 1 year or more prior to harvest can increase the fine root density in the root ball 4- to 6-fold, compared to unpruned plants (23,71). Since the soil volume and associated moisture near these additional roots is unchanged, increased root density in the root ball may not be any advantage unless the root ball soil moisture is replenished frequently.

In-ground fabric containers (Figure 4) were developed to restrict the lateral spread of roots, promoting a root system with more roots and smaller roots than those inside a traditional field-grown root ball. This, in turn, would lead to more rapid root regeneration after transplanting (73). The response of trees to the fabric container has been variable. There are reports of increased, unchanged (31) and decreased (7) root weight compared to a field-grown root ball. There is little evidence linking increased root dry weight within the root ball of fabric container with reduced stress or increased growth following

---

Gary W. Watson is a Root System Research Biologist with The Morton Arboretum, Lisle, Illinois 60532 USA.

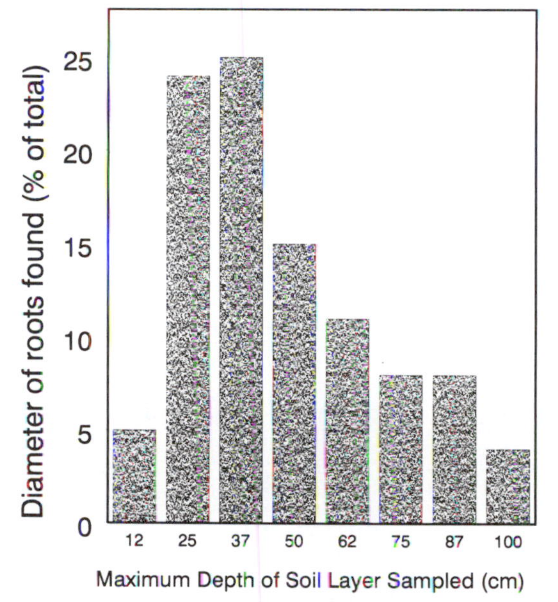

Figure 1 — The roots of 7 - 10 cm caliper shade trees growing in the nursery are most concentrated 13-38 cm below the surface.

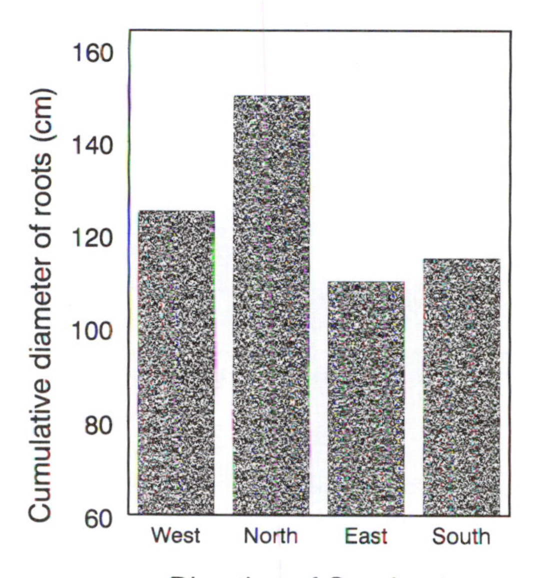

Figure 2 — The root system of 7 - 10 cm caliper shade trees growing in the nursery is more well-developed on the north side.

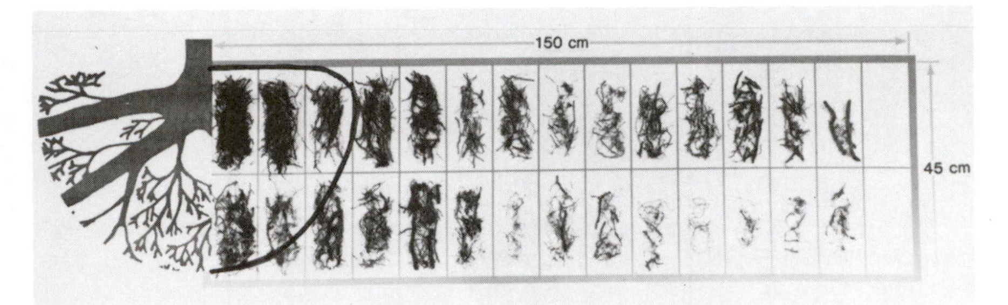

Figure 3 — Most of the root system is left in the nursery when field grown trees are harvested.

Figure 4 — In-ground fabric containers were designed to restrict the spread of the root system and to facilitate harvest, but their performance has been inconsistent.

transplanting. Root regeneration can be greater, the same (29) or lower (14) on trees grown in fabric containers.

Roots often circle along the inside of a smooth-sided container. Roots deflected by container walls can cause root deformations which may lead to long-term tree growth problems (46). Techniques that have been tested to disrupt these circling roots include variations of root ball slicing, teasing roots away from the periphery of the root ball, and striking the root ball against a concrete surface to loosen roots from the medium (6,60). None of these techniques improves woody plant root growth after transplanting and many of them reduced growth. Long-term effects of these treatments on the root system were not reported.

Formation of New Roots

Root Initiation

When roots are cut during transplanting, the majority of the regenerated roots emerge from immediately behind the injured surface of the roots that were severed during transplanting (Figure 5) (30,67,77). New roots curve abruptly as they emerge until they are growing approximately the same direction as the root that was severed. Preformed lateral roots angle only slightly or are at right angles.

Root regeneration characteristics are species dependent. Intact root tips of relatively easy-to-transplant green ash, (*Fraxinus pennsylvanica*) began elongation within 9 days of transplanting and adventitious roots were formed within 17 days (1). Adventitious roots of more difficult-to-transplant red oak (*Quercus rubra*) were not formed until 24-49 days after transplanting (57).

Longevity of Regenerated Roots

Usually, multiple roots are produced from the end of the severed root. Many of these live for only a short time. Roots with the largest diameters at the point of emergence become dominant (30). Eventually the smaller roots may die leaving little or no trace (Figure 6) (62). In a study of Colorado spruce (*Picea pungens*), the fate of regenerated roots was dependent on the diameter and age of the severed root. When large roots were severed, almost none of the roots regenerated near the injured surface persisted for five years. In contrast, when small roots were severed, the dominant regenerated root from each of these smaller severed roots grew vigorously, collectively forming the new structural root system (Figure 7).

Figure 5 — When roots are cut during transplanting, the majority of the regenerated roots emerge very near the severed root ends at the edge of the root ball.

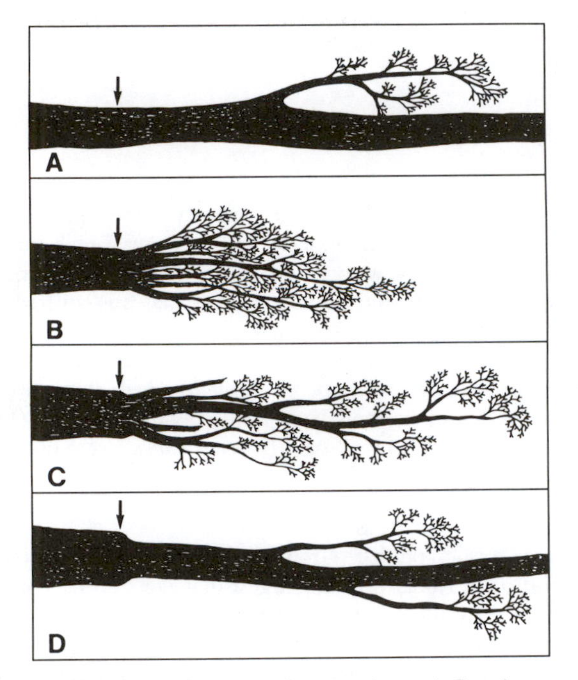

Figure 6 — Stages in replacement of a severed root. A. Root is severed at the arrow. B. Initially, many small roots are regenerated from the callus collar at the severed end. C. One root becomes dominant and continues to elongate while others remain stagnant or begin to die. D. Eventually, only a single root remains in place of the original root.

Use of Auxins to Increase Root Regeneration

Auxin applications increase the number of roots regenerated (33,40,41,51,64), but increase time to first root regeneration, and decrease root elongation rate (56). High concentrations of indole butyric acid inhibit regrowth (32). The most widely used methods of auxin application are root dips or sprays (33,55). Napthaleneacetic acid has also been used successfully as a backfill and root ball soil drench to increase the number of regenerated roots (63). As in the case of root pruning, this increase in root density may not improve survival and growth unless the more rapidly depleted soil moisture is replenished frequently. Use of hormones which increase root elongation could hasten the growth of roots into a larger soil volume with additional moisture available. More research is needed.

Factors Affecting New Root Growth and Development

Rate of Elongation

In the temperate climate of the northern United States, new roots elongate at a rate of approximately 30-60 cm/yr (10,17). In subtropical climate of Florida, root elongation rates from 60 to 110 cm/year have been reported (18,20). Live oak roots are capable of growing at a rate of 300 cm per year in Florida

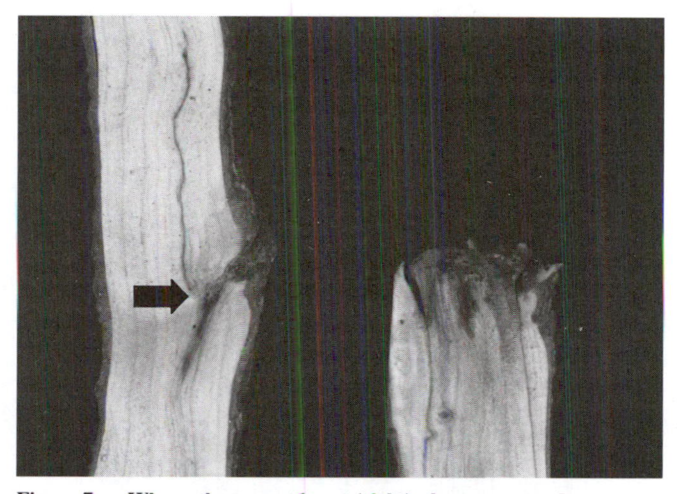

Figure 7 — When a large root is cut (right), the regenerated roots may not persist for long. When a small root is cut (left, at arrow), at least one of the regenerated roots usually develops into a major woody root of the new root system.

Soil Characteristics

Root branching is more prevalent in rich, porous soils. In compacted soil, when root growth is restricted by low soil oxygen supply and mechanical impedance, root development is often limited to soil cracks where aeration is adequate and resistance to penetration is minimal (19). Roots in compacted soil may be redirected up toward the soil surface where resistance to penetration is lower (22,70). In these compacted soils, only roots growing from the top portion of root ball, or those reaching the soil surface after being redirected are likely to develop as a significant part of the permanent structural root system (18,22). On poorly drained sites, the roots in the lower portion of the root ball can be killed. Trees planted in compacted urban soil perform best if root balls are placed slightly above existing grade (22,47).

Seasonal Trends

The period of maximum growth of roots is governed by soil moisture, soil temperature, stage of shoot development and other factors. On established trees, root growth often peaks in very early spring and again in early fall, before and after active shoot growth. Transplanted trees may not be subject to the same physiological control mechanisms. Each species has a characteristic shoot:root ratio (34). When the ratio is disturbed, plants respond by redirecting assimilates to replace the removed parts. This was demonstrated when trees were transplanted at four different times from early spring to fall. Root regeneration was greatest when trees were transplanted in mid-May (Figure 8), during the period of most active shoot growth and normally low root growth (67).

Apical Dominance

Woody lateral roots often grow very slowly, not often becoming part of the structural root system, but serving to facilitate the distribution of non-woody root tips throughout

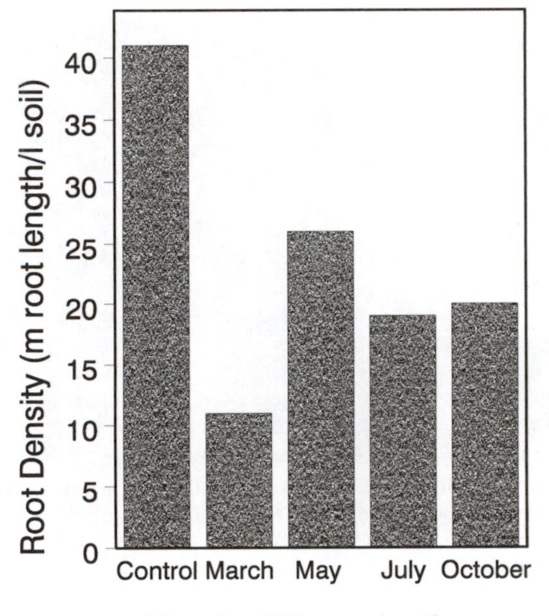

Figure 8. — Trees produced the greatest root growth in the first year when they were transplanted in May, even though root growth of established trees can be minimal during this time when shoot growth is active.

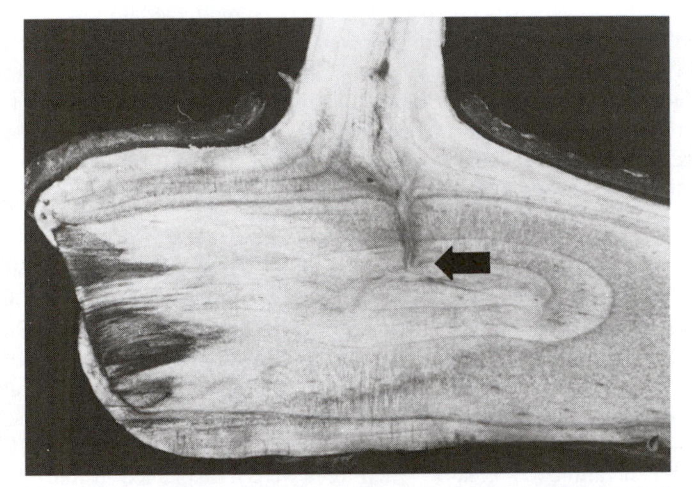

Figure 9 — Lateral root growth is increased by severing the terminal root. The lateral root originated in the second growth ring (arrow), but its diameter increase was very small for three years until after the terminal root was cut and annual rings began to increase in width.

the soil. When a large terminal root is cut during transplanting, growth of the laterals often increases markedly (Figure 9) (23). These roots may then become part of the permanent structural root system. As these roots continue to increase in diameter, difficulties may arise if they are perpendicular to the root from which they originated and lie across the base of the trunk or other roots. Roots can girdle the trunk or other roots as they enlarge (68). These girdling roots restrict secondary growth, reduce the flow in the vascular system and are associated with decline.

Planting and Maintenance Considerations

Tree planting on urban sites poses many challenges. The soils are often disturbed, alkaline, compacted, high in clay content, and poor in structure. Drainage, aeration, and water holding capacity are often very poor. These soils can change from waterlogged to very dry, in just a few days. When drainage is poor, greater success is often achieved by planting high, with up to one-third of the root ball above original grade. Larger planting holes (wider, not deeper) lead to greater transplanting success (8,43). We do not have a thorough understanding of the stresses involved in establishing trees on these sites.

Soil Amendments

Preparation of the planting hole is an extremely important factor in transplanting success (36). Reports on the growth response of roots when backfill soil differs from the site-soil are conflicting. Roots may have difficulty crossing the interface between the two different soils, restricting the roots to the planting hole (53). These studies on sandy agricultural soils provide little insight into urban situations with compacted, clayey soils. In another study on compacted clayey soil, roots were able to penetrate the interface between the backfill soils and the site-soil within the first season after transplanting, whether the backfill soil was similar to, or different from the site soil (70). Root density in the compacted subsoil outside of the hole was generally lower, but was attributed to the compacted, clay soil outside of the planting hole, rather than to difficulty crossing the interface. The degree and type of backfill soil amendment may be a major factor. Generally, the negative effects on root development were most noticeable when the amount of organic amendments added was very high (74).

Soil Moisture Utilization

Plant growth is reduced more often by water deficits than by any other factor (34). Water in soil that is not permeated by roots is largely unavailable for absorption by roots (35). Very limited water is available to the roots in the root ball. Progressive redevelopment of the root system of transplanted trees results in the exploitation of an ever-increasing soil volume containing a greater volume of available water (2,69).

Soil moisture uptake was measured on *Fraxinus pennsylvanica* trees that were recently planted in a compacted, clayey soil typical of many urban landscapes. For trees planted in April, transpirational water loss in June caused the root ball soil moisture tension (SMT) to increase by over -40 KPa in as little as two days. SMT frequently reached -50 KPa in the root ball while the SMT in the backfill soil, 3 cm from the root ball, remained under -10 KPa, indicating there was still little or no moisture uptake by roots outside the root ball more than two months after planting (Figure 10). Root growth can be inhibited at -50 KPa (4). During periods of high transpiration, newly transplanted trees may require watering as often as every one to three days, depending on temperature and rainfall, in order to be assured of avoiding drought induced inhibition of root growth.

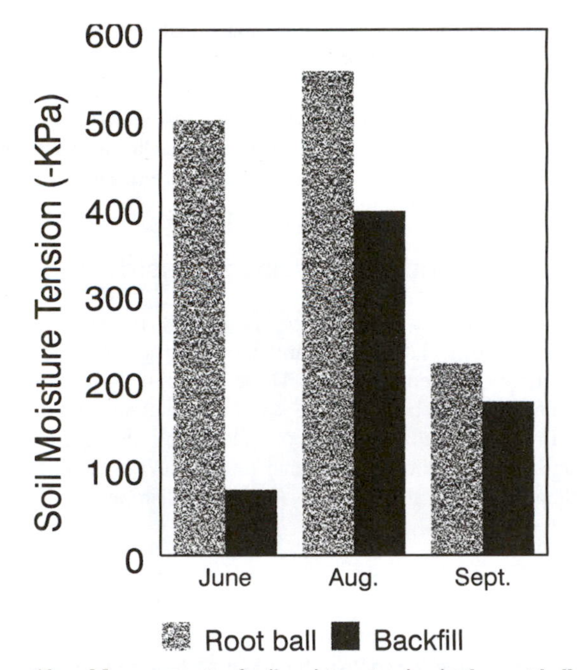

Figure 10 — Measurements of soil moisture tension in the root ball and adjacent to the root ball of 7 cm (2.5-inch) caliper trees transplanted in April show that there is little uptake of moisture outside of the root ball in June, two months after the trees were planted. At the same time, the root ball soil is very dry, as indicated by the high soil moisture tension measurement. Not until September is there similar uptake from the root ball and backfill soils.

It was not until early September (20 weeks after transplanting) that soil moisture was being depleted at the same rate both inside of the root ball and 3 cm outside of the root ball. It took until the end of the second season after planting (68 weeks) for water to be depleted at a similar rate 60 cm from the edge of the root ball. At this time, it was estimated that 9 times as much soil was exploited by the root system, and therefore up to 9 times as much moisture would be available for absorption by the root system. This water supply should be adequate for approximately two weeks without rainfall, compared to the two day supply of water during the first weeks after transplanting.

Container media are very coarse-textured, with a perched water table created by the bottom of the container. When these root balls are removed from the container and put into contact with landscape soils, up to 85 percent of available water can be lost to the finer-textured surrounding soil (9,45). However, since there is no root loss on transplanting, water stress may still be less severe than for B&B plants, as long as irrigation is provided regularly after planting.

Irrigation

Method of irrigation influences root distribution. In arid climates, drip irritation encouraged root growth only within an area 0.6 m from the emitters (38). In temperate climates, trickle irrigation caused no concentration of roots near the drip emitter. Root spread was similar to trees receiving overhead irrigation (16). There is often sufficient soil moisture for root growth well beyond the drip emitter in temperate climates (50).

Depth of wetting exerts a powerful influence on depth of rooting (11). Roots penetrate deeper into soil as the wetting depth increases. There can be a larger percentage of the root system in the shallow depths in soils kept continually moist than in those with periodic drying cycles.

Length of Establishment Period

The length of time required to fully reestablish the root system is dependent on the extent of the original root spread and the root elongation rate. Root spread generally increases with tree size. Annual elongation rate will vary with climate, and species, but not with tree size. Larger trees therefore require more annual increments of root growth to replace the root system, and have a longer establishment period (61). A smaller plant, which establishes its root system and resumes vigorous growth faster, can become larger than a tree planted at the same time that was initially much larger (Figure 11) (58). The time required for trees to regain pre-transplanting growth rate is approximately 1 year for each 2.5 cm caliper (64).

Fertilization

Soil fertility may also influence root growth (49). Numerous studies have shown that fertilization of newly planted trees is ineffective (59,78). Water stress is the most commonly limiting growth factor after transplanting.

Application of nutrients may stimulate root growth, but it can be very localized (13,27). If root growth is stimulated in one area, it can be at the expense of other portions of the root system (72). The increase in density is due to closer spacing between lateral roots at higher nutrient concentrations. Over fertilization with nitrogen can retard development of fine roots (42). High rates of nitrogen can encourage shoot growth at the expense of root growth (44,48,79).

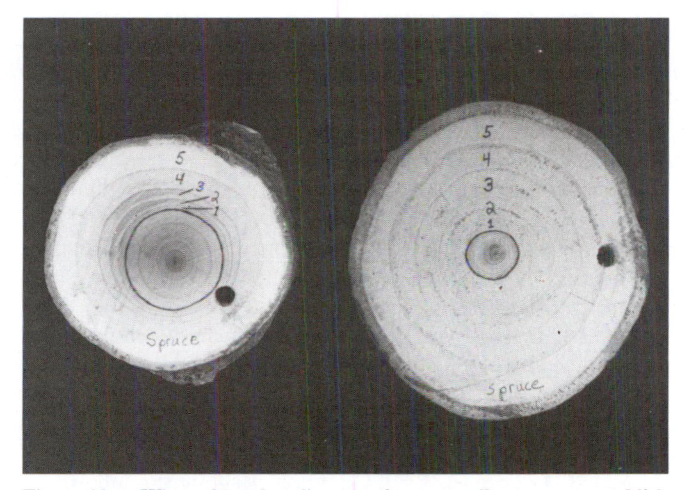

Figure 11 — When planted at the same time, a smaller tree can establish more quickly. The tree on the right, which was smaller in diameter when planted (indicated by circle), grew larger in 5 years.

Figure 12 — Root growth and crown development of recently transplanted trees improve significantly when mulched (tree on right). Unmulched tree is shown on the left.

Root Competition

Turfgrass is an aggressive competitor for soil moisture and nutrients. Most tree species that are used in the landscape evolved in the moist forest environment, without competition from grasses. Grasses are often associated with drier ecosystems. Grass is most competitive with tree roots if mowing is conducted infrequently. Turf roots are deeper and root density greater on unmowed or infrequently mowed turf (3,5). However, closely mowed grass has been associated with more rapid uptake of soil moisture. Competition is most pronounced in the top soil layers since roots of both plants proliferate in this well drained, well aerated, nutrient rich zone (15,25,65,75). Many landscapes receive applications of mulch to discourage grass and weed competition, buffer soil temperature and for other reasons (28). Root density can be increased significantly. Fine roots were 15 times as dense in mulched soil and were associated with more rapid establishment and improved growth of newly planted trees (Figure 12) (12,25).

Transplanting is not a natural process. Replacment of the roots that are lost takes time, and is influenced by many factors. Urban sites are often poor environments for root development. Until the root system is replaced, the tree is subject to stress. The more we learn about how roots react with these man-made soil environments, the more successfully we will be able to plant and grow trees in our cities.

Literature Cited

1. Arnold, M.A. and D.K. Struve. 1989. Green ash establishment following transplant. J. Amer. Soc. Hort. Sci. 114:591-595.
2. Barnett, D. 1988. Root growth and water use by newly transplanted woody landscape

plants. Public Garden (April) 23-25.
3. Beard, J.B. and W.H. Danial. 1965. Effect of temperature and cutting on the growth of creeping bentgrass roots. Agron. J. 57:249-250.
4. Bevington, K.B. and W.S. Castle. 1985. Annual root growth pattern of young citrus trees in relation to shoot growth, soil temperature, and soil water content. Amer. Soc. Hort. Sci. 110:840-845.
5. Biswell, J.J. and J.E. Weaver. 1933. Effect of frequent clipping on the development of roots and tops of grasses in prairie sod. Ecology 14: 368-390.
6. Blessing, S.C. and M.N. Dana. 1988. Post-transplant root system expansion in *Juniperus chinensis* as influenced by production system, mechanical root disruption, and soil type. J. Environ. Hort. 5:155-158.
7. Chong, C., G.P. Lumis, R.A. Cline and H.J. Reissmann. 1987. Growth and chemical composition of *Populus deltoides* x *nigra* grown in Field-Grow fabric containers. J. Environ. Hort. 5:45-48.
8. Corley, W.L. 1984. Soil amendments at planting. J. Environ. Hort. 2:27-30.
9. Costello, L. and J.L. Paul. 1975. Moisture relations in transplanted container plants. HortSci. 10:371-372.
10. Coutts, M.P. 1983. Development of the structural root system of Sitka spruce. Forestry 56:1-16.
11. Cullen, P.W., A.K. Turner and J.H. Wilson. 1972. The effect of irrigation depth on root growth of some pasture species. Plant and Soil 37:345-352.
12. Davies, R.J. 1987. Trees & weeds. Forestry Commission Handbook 2, Her Majesty's Stationery Office, London.
13. Eissenstat, D.M. and M.M. Caldwell. 1988. Seasonal timing of root growth in favorable microsites. Ecology 69:870-873.
14. Fuller, D.L. and W.A. Meadows. 1988. Root and top growth response of five woody ornamental species to fabric Field-Grow containers, bed height and trickle irrigation. Proc. S. Nurs. Assoc. Res. Conf. 34:148-151.
15. Gale, M.R. and D.F. Grigal. 1987. Vertical root distribution of northern tree species in relation to successional status. Can. J. For. Res. 17:829-834.
16. Gilman, E.F. 1988. Predicting root spread from trunk diameter and branch spread. J. Arboric. 14:85-89.
17. Gilman, E.F. 1988. Tree root spread in relation to branch dripline and harvestable root ball. HortSci. 23:351-353.
18. Gilman, E.F. 1989. Plant form in relation to root spread. J. Environ. Hort.7:88-90.
19. Gilman, E.F. 1990. Tree root growth and development. I. Form, spread, depth and periodicity. J. Environ. Hort. 8:215-220.
20. Gilman, E.F. 1990. Tree root growth and development. II. Response to culture, management and planting. J. Environ. Hort. 8:220-227.
21. Gilman, E.F. and M.E. Kane. 1990. Root growth of red maple following planting from containers in two different soils. HortSci. 25:527-528.
22. Gilman, E.F., I.A. Leone and F.B. Flower. 1987. Effect of soil compaction and oxygen content on vertical and horizontal root distribution. J. Environ. Hort. 5:33-36.
23. Gilman, E.F. and T.H. Yeager. 1987. Root pruning *Quercus virginiana* to promote a compact root system. Proc. S. Nurs. Assoc. Res. Conf. 32:340-342.
24. Gilman, E.F. and T.H. Yeager. 1988. Root initiation in root-pruned hardwoods. HortSci. 23:775.
25. Green, T.L. and G.W. Watson. 1989. Effects of turfgrass and mulch on theestablishment and growth of bare-root sugar maples. J. Arboric. 15:268-272.
26. Greenwood, D.J. 1968. Effect of oxygen distribution in the soil on plant growth. p. 202-

221. **In** W.J. Wittington (ed.), Root Growth, Plenum Press, New York.

27. Hackett, C. 1972. A method of applying nutrients locally to roots under controlled conditions, and some morphological effects of locally applied nitrate on the branching of wheat roots. Austral. J. Biol. Sci. 25:1169-1180.
28. Harris, R.W. 1992. Arboriculture: Integrated management of landscape trees, shrubs and vines. Prentice Hall, Englewood Cliffs, New Jersey.
29. Harris, R.W. and E.F. Gilman. 1991. Production method affects growth and root regeneration of leyland cypress, laurel oak and slash pine. J. Arboriculture 17:64-69.
30. Horsley, S.B. 1971. Root tip injury and the development of the paper birch root system. For. Sci. 17:341-348.
31. Ingram, D.L., U. Yadav and C.A. Neal. 1987. Production system comparisons for selected woody plants in Florida. HortSci. 22:1285-1287.
32. Kelly, R.J. and B.C. Moser. 1983. Root regeneration of *Liriodendron tulipifera* in response to auxin, stem pruning, and environmental conditions. J. Amer. Soc. Hort. Sci. 108:1085-1090.
33. Kling, G.J. 1984. Root regeneration techniques. Proc. Intl. Plant Prop. Soc. 34:618-627.
34. Kramer, P.J. 1983. Water relations of plants. Academic Press, New York.
35. Kozlowski, T.T. 1987. Soil moisture and absorption of water by tree roots. J. Arboric. 13:39-46.
36. Kozlowski, T.T. and W.J. Davies. 1975. Control of water balance in transplanted trees. J. Arboric. 1:1-10.
37. Kramer, P.J. and T.T. Kozlowski. 1979. Physiology of woody plants. Academic Press, N.Y.
38. Levin, I., R. Assaf and B. Bravdo. 1980. Irrigation water status and nutrient uptake in an apple orchard. p. 225-264. D. Atkinson, J.E. Jackson, R.O. Sharples, and W.M. Waller (eds.). **In** The mineral nutrition of fruit trees. Butterworths Borough Green, U.K.
39. Leyton, L. and L.Z. Roussear. 1958. Root growth of tree seedlings in relation to aeration. p. 467-475. **In** K. Thimann (ed.) The Physiology of Forest Trees.
40. Lumis, G.P. 1982. Stimulating root regeneration of landscape-size red oak with auxin root sprays. J. Arboric. 8:325-326.
41. Magley, S.B., and D.K. Struve. 1983. Effects of three transplant methods on survival, growth and root regeneration of caliper pin oaks. J. Environ. Hort. 1:59-62.
42. May, L.G., F.H. Chapman and D. Aspinall. 1964. Quantitative studies of root development. I. The influence of nutrient concentration. Austral. J. Biol. Sci. 18:25-35.
43. Miller, A.N., P.B. Lombard, M.N. Westwood and R.L. Stebbins. 1990. Tree and fruit growth of 'Napoleon' cherry in response to rootstock and planting method. HortSci. 25:176-178.
44. Nambiar, E.K.S. 1980. Root configuration and root regeneration in Pinus radiata seedlings. N.Z. J. For. Sci. 10:249-263.
45. Nelms, L.R. and L.A. Spomer. 1983. Water retention of container soil transplanted into ground beds. HortSci. 18:863-866.
46. Nichols, T.J. and A.A. Alm. 1983. Root development of container-reared, nursery-grown, and naturally regenerated pine seedlings. Can. J. Bot. Res. 13:239-245.
47. Patterson, J.C. 1976. Soil compaction and its effect on urban vegetation. p. 91-102. **In** F. Santamour, J.D. Gerhold and S. Little (eds.). Better trees for metropolitan landscapes. Sympos. Proc. USDA For. Serv. Gen.Tech.Rep. NE-22.
48. Pham, C.H., H.G. Halverson and G.M. Heisler. 1978. Red maple (Acer rubrum L.) growth and foliar nutrient responses to soil fertility level and water regime. U.S. Forest Service Research Paper NE-412.
49. Philipson, J.J. and M.P. Coutts. 1977. The influence of mineral nutrition on the root

development of trees, II. The effect of specific nutrient elements on the growth of individual roots of sitka spruce. J. Exp. Bot. 28:864-871.
50. Ponder, H.G. and A.L. Kenworthy. 1976. Trickle irrigation of shade trees growing in the nursery. J. Amer. Soc. Hort. Soc. 101:104-107.
51. Prager, C.M. and G.P. Lumis. 1983. IBA and some IBA-synergist increases of root regeneration of landscape-size seedling trees J. Arboric. 9:117-123.
52. Rogers, W.S. and G.C. Head. 1968. Factors affecting the distribution and growth of roots of perennial woody species. **In** W.H. Whittington (ed.) Root Growth, Plenum Press, New York.
53. Schulte, J.R. and C.E. Whitcomb. 1975. Effects of soil amendments and fertilizer levels on the establishment of silver maple. J. Arboric. 1:192-195.
54. Stout, B.B. 1956. Studies of root systems of deciduous trees. Black Rock For. Bull. No. 15. Harvard University Printing Office. Cambridge, Massachusetts. 45pp.
55. Struve, D.K., R.D. Kelly, and B.C. Moser. 1983. Promotion of root regeneration of difficult-to transplant species. Proc. Intl. Prop. Soc. 33:433-439.
56. Struve, D.K. and B.C. Moser. 1984. Auxin effects on root regeneration of scarlet oak seedlings. J. Amer. Soc. Hort. Sci. 109:91-95.
57. Struve, D.K. and W.T. Rhodus. 1988. Phenyl indole-3-thiobutyrate increases growth of transplanted 1-0 red oak. Can. J. For. Res. 18:131-134.
58. van de Werken, H. 1970. Establishing shade trees in lawns. Tennessee Farm and Home Science Progress Report No. 73, pp 5-7.
59. van de Werken, H. 1981. Fertilization and other factors enhancing the growth rate of young shade trees. J. Arboric. 7:33-37.
60. Wade, G.L. and G.E. Smith. 1985. Effect of root disturbance on establishment of container grown Ilex crenata 'Compacta' in the landscape. Proc. S. Nurs. Assoc. Res. Conf. 30:110-111.
61. Watson, G.W. 1985. Tree size affects root regeneration and top growth after transplanting. J. Arboric. 11:37-40.
62. Watson G. W. 1986. Cultural practices can influence root development for better transplanting success. J. Environ. Hort. 4:32-34.
63. Watson, G.W. 1987. Are auxins practical for B&B trees? Amer. Nurs. 6(4):183.
64. Watson, G.W. 1987. The relationship of root growth and tree vigour following tree root development. J. Arboric. 14:200-203.
65. Watson, G.W. 1988. Organic mulch and grass competition influence tree root development. J. Arboric. 14:200-203.
66. Watson G.W. and E.B. Himelick. 1982. Root distribution of nursery trees and its relation to transplanting success. J. Arboric. 8:225-229.
67. Watson, G.W. and E.B. Himelick. 1982. Seasonal variation in root regeneration of transplanted trees. J. Arboric. 8:305-310.
68. Watson, G.W., S. Clark and K. Johnson. 1990. Formation of girdling roots. J. Arboric. 16:197-202.
69. Watson, G.W. and G. Kupkowski. 1991. Soil moisture uptake by green ash trees after transplanting. J. Environ. Hort. 9:227-230.
70. Watson, G.W., G.K. Kupkowski and K.G. von der Heide-Spravka. 1992. The effect of backfill soil texture and planting hole shape on root regeneration of transplanted green ash. J. Arboric. 18:124-129.
71. Watson G.W. and T.D. Sydnor. 1987. The effect of root pruning on the root system of nursery trees. J. Arboric. 13:126-130.
72. Weller, F. 1966. Horizontal distribution of absorbing roots and the utilization of fertilizers in apple orchards. Erwobstbsobstbau 8:181-184.

73. Whitcomb, C.E. 1986. Fabric Field-Grow containers enhance root growth. Amer. Nurs. 163:49-52.
74. Whitcomb, C.E. 1987. Establishment & maintenance of landscape plants. Lacebark. Stillwater, Oklahoma.
75. Whitcomb, C.E. and E.C. Roberts. 1973. Competition between established tree roots and newly seeded Kentucky bluegrass. Agron. J. 65:126-129.
76. White, G.C. and R.I.C. Halloway. 1967. The influence of Simizine on a straw mulch on the establishment of apple trees in grassed and cultivated soil. J. Amer. Soc. Hort. Sci. 42:377-389.
77. Wilcox, H. 1955. Regeneration of injured root systems in Noble fir. Bot. Gaz 116:211-234.
78. Wright, R.D. and E.B. Hale. 1983. Growth of three shade tree genera as influenced by irrigation and nitrogen rates. J. Environ. Hort. 1:5-6.
79. Yeager, T.H. and R.D. Wright. 1981. Influence of nitrogen and phosphorus on shoot:root ratio of *llex crenata* Thumb. 'Helleri'. HortSci. 16:564-565.

Establishing Trees in the Landscape

Edward F. Gilman

Tree establishment does not begin at the planting site. It begins at the nursery by picking appropriate nursery stock. Suitable nursery stock must be chosen based on planting site conditions and intended after care. If not, establishment efforts at the planting site can be futile. Several factors affect the suitability of a particular type of nursery stock to the planting site. They include tree size, root ball characteristics, and nursery practices.

Trees require special care until they are established. The time required to establish a tree depends on the climate and irrigation management after planting.

Largest recommended tree size at transplanting should be governed by irrigation capabilities after planting, and climate (Figure 1). If irrigation can not be provided for the recommended period after planting a certain size tree, choose smaller nursery stock to ensure survival and good growth.

Root balls of any shape can be planted in well-drained soil. Low-profile root balls are probably better suited than traditionally-shaped nursery stock for planting in poorly-drained and compacted sites (Figure 2). Low-profile root balls come from low-profile containers or from a field nursery with a compacted subsoil or high water table. The root ball shape from a low-profile container resembles the ultimate shape of a tree root system, i.e. shallow and wide (see front cover). Roots in the deeper part of a traditionally-shaped root ball could die if they are submerged in water or placed in a compacted soil with a low oxygen content. This could hinder or prevent establishment.

To ensure greater tree transplant survival, it is essential to choose trees grown in a nursery production system best suited for the characteristics and anticipated management practices at the planting site. The choice is simple for a well-drained site. If trees receive regular irrigation after planting, trees from all production systems perform almost equally well (Table 1). Professional judgement based on the following discussion is crucial in less hospitable sites.

Nursery Production Methods Impact Transplanting

Comparing Methods

The weight of scientific evidence shows that root balls of field-grown trees are similar to those grown in fabric containers except that fabric-grown root balls are about 50% smaller and are easier to handle (Figure 3). Recent research demonstrates that about the same percentage of roots are harvested from both production methods (Figure 4).

Edward F. Gilman is with the Department of Ornamental Horticulture, University of Florida, Gainesville, FL 32611.

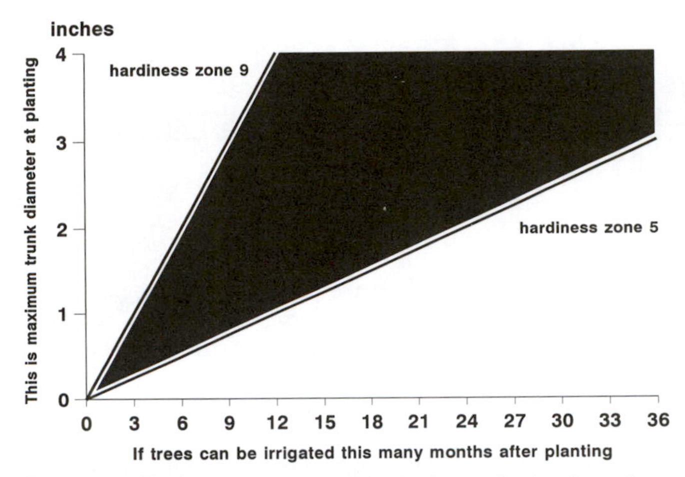

Figure 1 — The largest recommended tree size at planting depends on the climate and the length of time irrigation can be provided after planting.

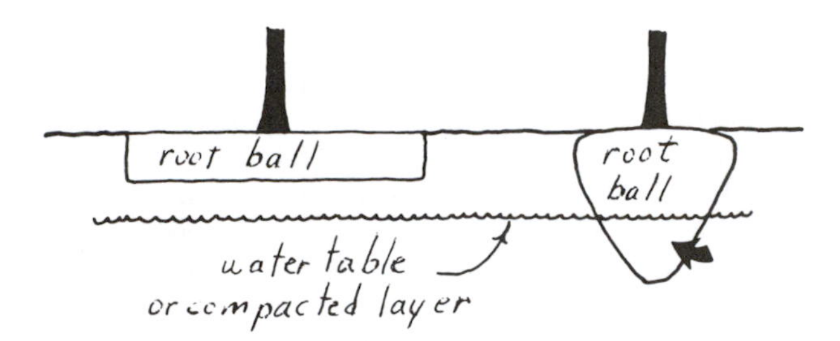

Figure 2 — Trees with a low-profile root ball are well suited for planting in compacted or wet sites (left). Roots at the bottom of conventional root balls (thick arrow) may die when planted into poorly-drained or compacted sites (right).

Because of the smaller root ball, there is less reserve water in the fabric container root ball than in the larger-sized root ball of the field-grown tree (8). Combined with the dense root system, this makes trees from fabric containers more sensitive to desiccation immediately after digging than trees grown directly in the field soil (1,9). Nursery operators can usually provide the irrigation needed to prevent desiccation. A container root ball has many times more fine roots than that of a similarly-sized tree harvested from a field nursery (Figure 5). This dries the root ball quickly after planting and helps make container trees more sensitive to drought stress after planting.

Trees from fabric containers are as viable as those grown directly in the field if they are handled and irrigated properly after digging from the nursery. However, be sure to purchase field and fabric container-grown trees that have been dug at least several weeks earlier in order to ensure that the tree will survive the shock from digging. Until they are

Table 1. Choosing trees by production method

Production* method	Root ball weight	Need staking?	IF: Irrigation after planting is:	THEN: Root regeneration and trunk growth is:	AND: Survival is:
Above ground container	Light	Frequently	Frequent	Good to excellent	Excellent
			Infrequent	Fair to good	Fair
Fabric container	Light to moderately heavy	Yes	Frequent	Excellent	Excellent
			Infrequent	Good**	Good**
B&B	Heavy	Sometimes	Frequent	Excellent	Excellent
			Infrequent	Good**	Good**

*There is little research comparing bare-root trees with trees produced by other methods. There is little research comparing field-grown trees potted into containers with trees produced by other methods.

**Survival is good for trees hardened off in the nursery by pre-digging or root pruning several weeks to several months before planting. Survival and growth are only fair for fabric-container and B&B trees dug and immediately transplanted in the landscape without frequent irrigation.

Figure 3 — Root balls from trees grown in fabric containers (left) are identical to those grown directly in field soil (right) except that roots from the fabric container are concentrated in half the soil volume.

established in the landscape, trees harvested from fabric containers will require more frequent irrigation than those from a field nursery (1,9).

Root regeneration on container-grown trees lagged behind that of field-grown trees (Figure 6). Further study showed that this probably was related to the greater water stress experienced on container-grown trees following planting, unless they received daily irrigation. Although this stress can be overcome by frequent irrigation after planting, this is not currently practiced in the industry. Therefore, few trees planted from containers ever receive sufficient water.

It is not surprising to find that field-grown trees regenerate roots quicker than container-grown trees since a large portion of the root system is removed from field-grown trees at transplanting. The field-grown tree slows its shoot growth so it can replace its severed root system. There is no incentive for the tree planted from a container to quickly produce roots in the backfill soil.

Despite differences in root regeneration after transplanting, shoot growth on con-

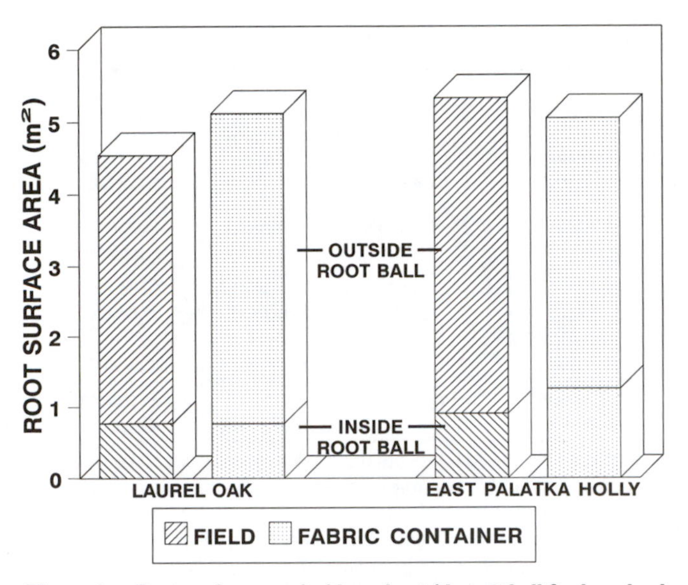

Figure 4 — Root surface area inside and outside root ball for laurel oak and East Palatka holly grown in field soil and in a fabric container.

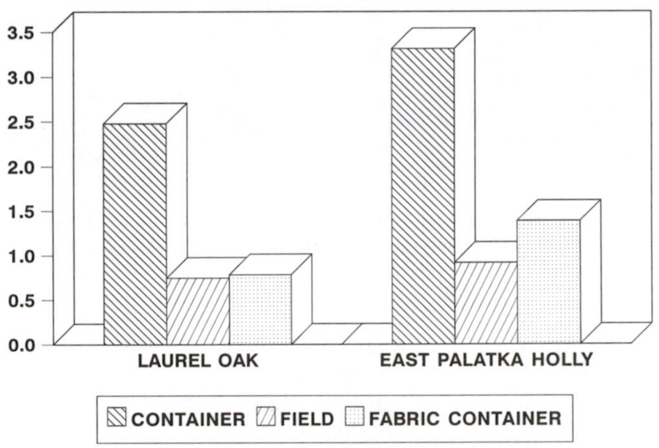

Figure 5 — Root surface area within the root ball of laurel oak and East Palatka holly grown in containers, in the field and in fabric containers.

tainer-grown trees appears to be similar to that on hardened-off field-grown trees provided irrigation is supplied until trees are established. Recently transplanted trees that have not been hardened-off from a field or fabric-container nursery are more susceptible to desiccation than trees from plastic containers (9).

In a recent study, live oak trees grown in containers treated with copper on the inside

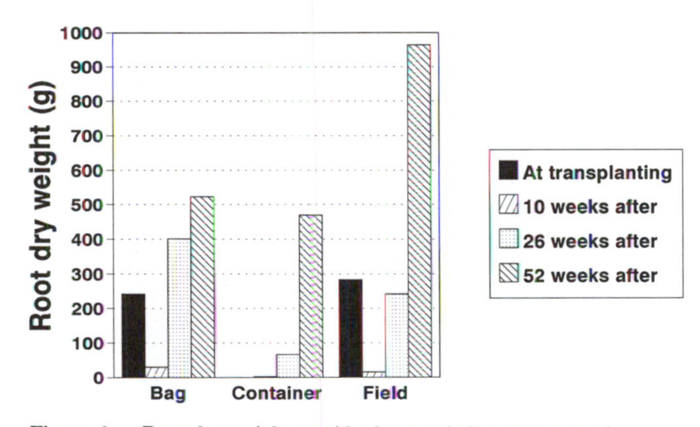

Figure 6 — Root dry weight outside the root ball at transplanting, then 10, 26, and 52 weeks after transplanting laurel oak from plastic containers, fabric containers (bag) and from the field.

surface were more stressed than those grown in conventional containers in the first three weeks after planting to the landscape unless they were irrigated daily. With daily irrigation there was no difference in stress levels. When irrigation was discontinued four to six weeks after planting, trees from copper containers were slightly less stressed than those from conventional containers. The reason for this is unclear since root regeneration was similar for trees planted from both types of containers (Figure 7).

Bare Root

Bare root trees are the least expensive. Unlike balled-in-burlap trees, a large portion of the root system can be harvested from the nursery and, unlike all other harvesting methods, the root system can be thoroughly inspected for defects.

Bare root trees are very sensitive to drying. If provisions are made to keep roots in the shade and moist during storage and transport and they are regularly irrigated after planting, they should perform as well as trees from other production methods. Bare root trees are only available in a limited range of sizes.

Trees as large as 3-inch caliper (live oak) can be successfully transplanted bare root in mid summer. Success depends on species and time of year.

Balled-in-burlap

Tree production practices in the field nursery impact transplantability. Trees receiving drip irrigation at the base of the trunk during production often have a concentration of fine roots in the root ball (Figure 8) (6). In some species fine root growth can also be stimulated close to the trunk by making fertilizer applications within the area of the future root ball. Another study casts doubts. Concentrating the irrigation or fertilizer close to the trunk after trees were in the ground for two years did not influence fine root growth (unpublished). Apparently, to increase fine-roots near the trunk, irrigation and fertilizer application to the root ball must begin soon after the tree is planted in the nursery.

Root pruning can impact transplantability. Trees dug and stored or trees root pruned several weeks or months prior to transplanting were less stressed after planting than freshly dug trees. On the other hand, if irrigation is provided regularly until trees are

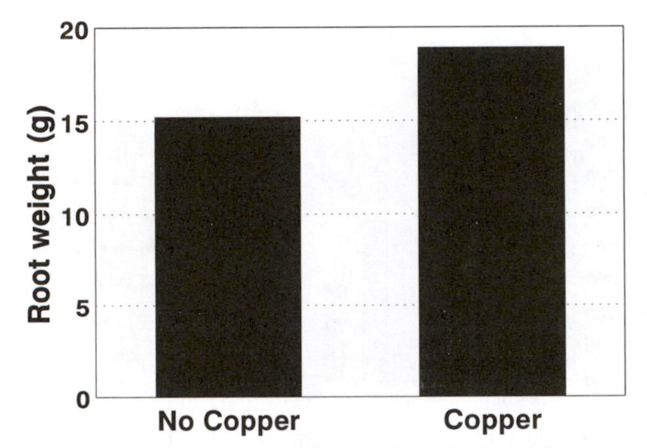

Figure 7 — Regenerated root weight from live oak planted from copper treated containers and non-treated containers.

established, root pruning prior to transplanting had no impact on survival or post-transplant growth on trees with a trunk caliper less than 4-inches (4).

Most trees grown in the field are dug and wrapped in burlap when harvested. Treated or untreated natural burlap usually decomposes in the soil and does not have to be completely removed at planting. However, professionals remove all burlap from the top of the ball so the sun does not wick moisture from the soil. All synthetic burlap should be removed from the root ball before backfilling. There are many reports that synthetic burlap strangles roots as they grow larger.

Synthetic burlap cannot be removed from a root ball tightly secured in a wire basket. To ensure continued health of the tree, remove the wire basket from the top of the root ball and cut away synthetic burlap. Natural burlap can remain in place.

Tree Spade

Trees transplanted with a tree spade generally respond like trees balled-in-burlap with the following exception. If the planting hole is dug with a tree spade in clay soil and the sides of the hole are glazed, then some roots could have trouble growing into the landscape soil. Enlarge the hole before planting so roots can penetrate loosened backfill soil. Lower the spade with the root ball into the hole and partially fill in around the spade with loosened backfill soil. Firm the soil and add water to settle.

Trees moved with a spade into loamy or sandy soil can usually be planted into a hole dug with the same size spade. The tree often ends up a little higher than it was in the nursery. This is much better than planting too deeply.

Hardened-off Trees

Trees dug several weeks or months prior to shipping to the landscape site are said to be `hardened off.' Freshly dug trees are not `hardened off'. Roots regenerate and grow within the root ball and the tree usually drops leaves during the `hardening off' period. `Hardened off' trees are more tolerant of dry soil after transplanting than are freshly dug trees.

Containers

In-ground Containers. Some nursery operators are beginning to grow trees in a

Figure 8 — Trees receiving irrigation only near the base of the trunk (center row) had more fine roots in the root ball than those receiving no irrigation (left row) or those receiving irrigation over a broader area (right row).

number of plastic containers sunk into the ground. Like standard, above-ground containers these are filled with artificial media and have a number of specially designed drainage holes along the sides and bottom of the container. Although there are no published comparative tests, trees grown in this fashion should transplant similarly to those grown in traditional, above-ground, rigid-plastic containers.

Above-ground Containers. Trees are grown in a variety of above-ground containers in the nurseries. Since root balls are designed to dry quickly to prevent root rot in the nursery, in the eastern U.S. daily irrigation is required in the summer following planting. In drier climates, container media are designed to hold moisture, and containers may not need daily irrigation.

Physical properties in the container maintain a perched water table at the bottom of the container. The water table disappears when the root ball is removed from the container and planted in the landscape. In addition, root density in the container is much greater than in B&B root balls. Consequently, the root ball dries quicker in the landscape than it did in the container (2). To maintain optimum growth after planting, container trees should be watered at least as often as they were in the nursery. In well-drained soil, daily irrigation will be required for a number of weeks (1-gallon containers) or months (30-gallon containers or larger) after planting, especially for trees with trunk diameters over two inches. Irrigation can be tapered off as roots begin to grow out into the landscape soil. Even then be prepared to irrigate during a drought until trees are fully established.

Nursery operators choose among the various container types based on a variety of production concerns in the nursery. As far as we know, trees transplant equally well from all above-ground container production systems.

Collected Native Trees

Oaks can be collected from native stands and transplanted to landscape sites. Research on laurel oak in Florida indicates that root balls on such trees are similar to those on nursery-grown trees that are not root pruned. Collected trees have a slightly less fine-root mass than trees grown in the nursery (5). Collected trees grow slower after transplanting than trees from other production systems. In addition, they often have a thinner canopy for a period of time following transplanting.

Establishment Period

The establishment period lasts 3 (in USDA hardiness zone 9) to 12 (in USDA hardiness zone 5) months per inch of trunk diameter. For a two-inch diameter tree in zone 9, this translates into a 6-month establishment period. For a two-inch diameter tree in zone 5, this translates into a 24-month establishment period. Adequate irrigation promotes quick establishment. Trees need regular irrigation early to ensure survival and less as roots grow into surrounding landscape soil.

By the end of the establishment period in moist climates a tree has regenerated enough roots to keep it alive without supplemental irrigation. In the drier parts of the U.S., the irrigation system combined with rainfall can provide enough water for survival after establishment.

Irrigation

Trees transplanted from containers take longer to establish than field-grown trees. Recent research indicates that they could take one or two months longer per inch trunk diameter (1). The mortality rate for container-grown trees is higher than for hardened-off field-grown trees if irrigation is cut off too soon. Field-grown trees placed in containers until roots fill the media are likely to act more like hardened off field-grown trees.

For rapid growth, trees planted from containers require more frequent irrigation than they received in the nursery. Frequent irrigation provides more benefit than applying large volumes of water infrequently. This is in direct contrast to the recommendation for established trees. Occasional irrigation of established trees with large water volumes is considered better than light, frequent applications.

Gradually increase the area irrigated around the tree to accommodate root growth, especially in the drier climates. Roots grow about one-half (hardiness zone 5) to two (hardiness zone 9) inches per week during the first few years after planting (3,12).

Trees may survive and eventually become established with less frequent irrigation than outlined above, but you are taking a chance. If hot weather or drought were to occur before trees are established, trees could die back. Intensive irrigation management in the first several months after planting reduces the length of the stressful establishment period by promoting rapid root growth. Trees that regenerate roots rapidly establish quickly and are more tolerant of drought than those than are not established.

Trees without adequate irrigation during the establishment period may also develop a weak, multi-trunk habit. This is a response to tip die-back on the main trunk and branches. More research is needed in this area. If proper irrigation can not be provided following planting for the recommended period (Figure 1), consider planting smaller trees that establish quicker.

Mulching

Mulching recently transplanted trees increases tree growth after planting by reducing turf competition and enhancing tree root growth (7). Mulching should be encouraged. The only precaution is to keep it several inches away from the trunk so it will not rot the trunk.

Fertilization

Research on trees transplanted from field nurseries indicates that if there is a benefit to fertilizing at planting, it is small or undetectable (11). In no case has research indicated that fertilizer application at planting increased survival. Post-planting management efforts in the months following planting are better spent on irrigation than on fertilization.

There is little fertilization research on trees planted from containers. If trees can be irrigated regularly after planting, fertilizer spread soon after planting may enhance tree growth slightly. Since regularly-irrigated container trees suffer little transplant shock, they may respond to an early application of fertilizer sooner than trees transplanted from the field. This speculation has not been tested.

Nitrogen is the only element that occasionally gives a slight growth benefit (10). Potassium and phosphorus do not. There is no scientific evidence that other types of fertilizers, soil amendments or additives, natural or synthetic, help trees become established. Proper irrigation management is the key to success.

Literature Cited

1. Beeson, R. C. and E. F. Gilman. 1992. Diurnal water stress during landscape establishment of slash pine differs among three production methods. J. Arboric. 18:281-287.
2. Costello, L. and J. L. Paul. 1975. Moisture relations in transplanting container plants. HortScience 10:371-372.
3. Gilman, E. F. 1988. Tree root spread in relation to branch dripline and harvestable root ball. HortScience 23(2):351-353.
4. Gilman, E. F. 1992. Effects of root pruning prior to transplanting on establishment of southern magnolia in the landscape. J. Arboric. 18:197-200.
5. Gilman, E. F., R. C. Beeson and R. J. Black. 1992. Comparing root balls on laurel oak transplanted from the wild with those of nursery and container grown trees. J. Arboric. 18:124-129.
6. Gilman, E. F. and G. Knox. 1994. Micro-irrigation affects growth and root distribution of trees in fabric containers. HortTechnology (in press)
7. Green, T. L. and G. W. Watson. 1989. Effects of turfgrass and mulch on the establishment and growth of bare-root sugar maple. J. Arboric. 15:268-272.
8. Harris, J. R. and E. F. Gilman. 1992. Production system affects growth and root regeneration of leyland cypress, laurel oak and slash pine. J. Arboric. 17:64-69.
9. Harris, J. R. and E. F. Gilman. 1993. Production method affects growth and post transplanting establishment of East Palatka Holly. J. Amer. Soc. Hort. Sci. 118:194-200.
10. Neely, D., E. B. Himelick and W. R. Crowley. 1970. Fertilization of established trees: A report of field studies. Illinois Natural History Survey Bulletin 30(4):235-266.
11. Van de Werken, H. 1981. Fertilization and other factors enhancing the growth rate of young shade trees. J. Arboric. 7:33-37
12. Watson, G. W. 1985. Tree size affects root regeneration and top growth after transplanting. J. Arboric. 11:37-40.

Street Tree Establishment

Daniel K. Struve

Growth and physiology of transplanted trees, relative to untransplanted trees, are discussed. Transplant shock is initiated by massive root loss at harvest and is enhanced by the plant's developmental sequence after transplanting; leaves develop before roots regenerate. Transplant shock also contributes to maintenance problems such as formation of multiple leaders and heavy lower limb development. Container production systems that use copper-treated containers offer a means of rapidly producing difficult to transplant species while maintaining high root regeneration potential, thus reducing transplant shock. Site quality has a significant effect on transplant survival and regrowth. Early, limited results suggest that poor site quality can be easily improved at transplanting. Plant physiology can be altered up to three years after transplanting.

This paper discusses transplant establishment of northern temperate deciduous woody plants. The paper offers one definition of transplant shock, describes plant response to transplanting, discusses recent transplanting results of little-used species and suggests methods of improving transplant survival and regrowth. This paper is not an extensive review of root growth, root regeneration, transplanting or establishment of street trees. Examples are drawn primarily from the author's experience. The combined proceedings will offer a comprehensive review of the landscape below ground and its effects on tree growth.

Transplant shock is the period of slow growth following transplanting. It is characterized by limited shoot extension and sparse canopy development, the result of small individual leaf size and reduced leaf number, compared with established trees. Transplant shock may result in tree death. A good definition, but not fully adequate, of establishment is when the tree resumes pre-transplant growth rate. The limitation of this definition is that some trees may never become established. Site limitations may not support the pre-transplant growth rate achieved under the nursery environment.

Transplant shock is attributed to the massive loss of roots when plants are harvested (10. See also figures in 6). Root regeneration must occur for the transplant to survive and become established (9). The landscape (environment) below ground greatly influences the rate and degree of root regeneration and thus affects transplant survival and establishment. This paper begins by describing characteristic transplant shock symptoms and physiology which are a reflection of the quality of the below ground environment.

Daniel K. Struve is Associate Professor, Department of Horticulture, 2001 Fyffe Court, The Ohio State University, Columbus, OH 43210

Transplant Problems

Plant death is a common transplant problem. For instance, 1 to 1.25 in caliper bare root red oak have been planted twice at one Columbus, OH site. Of the 23 red oak planted, 20 are dead (87% mortality). The site is scheduled to be planted again this fall. Low survival is due to planting stock type, low site quality (curb lawn area with compacted clay soil) and minimal after care by the home owners.

If a tree survives transplanting, there are other long term expressions of transplant shock. Two expressions of transplant shock are large diameter lower limbs and multiple (co-dominate) leader development. Street trees often have large diameter lower limbs and co-dominate terminal shoots originating 6 to 8ft in height, about the height of the liners when transplanted. This form contrasts with the form of naturally regenerated (untransplanted) trees (Fig. 1A and B). For instance, silver maple (*Acer saccharinum*) can have an almost clear bole with few small diameter lower limbs (Fig. 2A and B). However, a specie's natural form can be lost at transplanting. The apical bud can be broken at or before transplanting or transplant stress can reduce apical dominance allowing lateral buds to grow with as much vigor as the terminal bud. Large diameter lower limbs are difficult-to-correct maintenance problems; limb removal leaves large pruning scars (Fig. 3). Research is needed to develop production systems where apical dominance is not lost at transplanting and/or to develop pruning methods that compensate for loss of

Figure 1 — A. Green ash planted in a curb lawn showing large diameter lower limb and multiple leader development approximately 8 feet above ground, about the height of the whip when planted. Large diameter lower limb and multiple leader development may be induced by loss of apical dominance attributed to transplant shock. B. The form of a naturally regenerated (untransplanted) green ash where large diameter lower limb development did not occur.

Figure 2 — A. Transplanted silver maple showing multipule leader development and B. the form of a naturally regenerated (untransplanted) silver maple. Note the abscence of large diameter lower limbs and a single bole. These two trees are within 50 yards of each other.

apical dominance when trees are transplanted.

Transplant shock is initiated by the massive root loss at harvest (10). It is further developed by the sequence of plant development after transplanting. In greenhouse studies, bud break precedes root regeneration (1,2). The time between bud break and first root regeneration may be as long as 33 days (2). Even when root regeneration occurs from intact root tips (container-grown stock) there can be a 14 day delay between bud break and first root elongation. When root regeneration is from new root initiation, longer periods between bud break and root regeneration occur. Typically, the greater the pruned root diameter, the slower is new root initiation (1,5). Thus, coarse-rooted species are difficult to transplant because most root regeneration is from new root initiation, first root regeneration occurs later and fewer roots are regenerated than in more fibrous rooted species (5).

Transplant Physiology and Growth

If bud break precedes root regeneration in transplanted plants, then the plant should be under water stress; a plant with a greatly reduced (pruned) suberized root system is less able to meet plant water requirements than a plant with a large unsuberized root system. Thus, transplanted plants should have reduced leaf surface area, shoot growth and photosynthetic gas exchange and more negative xylem water potentials than untransplanted plants. This hypothesis was tested with red oak seedlings under greenhouse conditions (8). One year old red oak seedlings were raised in three foot long containers.

Figure 3 — Pruning scar resulting from removal of a large diameter lower limb of a green ash.

Before bud break, half of the seedlings were removed from the containers, root pruned at six inches from the root collar and "transplanted" back into the containers. The other half of the seedlings were not root pruned and left undisturbed in the containers as untransplanted controls.

Under greenhouse conditions, the transplanted seedlings were not significantly different from untransplanted seedlings in new stem length, leaf number, end-of-experiment pre-dawn water potential, or net assimilation rate, stomatal conductance or transpiration on a per unit leaf area basis. Transplanted seedlings did have significantly less (45%) leaf surface area than untransplanted seedlings, 307 vs 557 cm^2. Under these experimental conditions, transplanted seedlings were not under as much stress as expected. Transplanted seedling adjusted to transplanting by reducing leaf surface area, thus reducing whole plant water use without altering per unit leaf area photosynthetic gas exchange.

These results were obtained under greenhouse conditions, conditions less stressful than outdoor conditions. It is possible that under less optimal conditions, transplanted plants are under more stress than untransplanted or established plants. This hypothesis was tested with 2 in caliper balled and burlaped shingle, chestnut and English oaks. Plants for the experiment were raised at the horticulture farm, spring dug and transplanted to sites on or near the Ohio State University campus. For all species, at least five plants were not dug to serve as "established" controls.

Shingle, chestnut and English oak all had reduced growth; shorter shoot length, reduced leaf number and caliper, up to two years after transplanting (Table 1). Average leaf size can recover within two years after transplanting if the transplanting site is modified with soil amendments; see the shingle and English oak two years after transplanting treatment in Table 1. The site modification practices will be discussed in a following section. In another experiment, 2" caliper balled and burlaped red oak were transplanted to two adjacent sites, distinguished by soil type. Plants were transplanted in fall 1989. The "fill" site had 6 to 8' of compacted clay subsoil layered over the native Kokomo silt clay soil type. The native soil type was a Kokomo silty clay with established sod; similar to the nursery soil type. For red oak planted on a poor quality site (Fill site condition), growth was reduced for three years (Table 2). Leaf size of transplanted trees on the native siol type was greater than untransplanted trees three years after transplanting, but photosynthesis (on a per leaf area basis) was less than one half the rate of untransplanted red oaks (Table 3).

Two years after transplanting, shingle and English oak had greater leaf size than untransplanted trees, but, as with red oak, photosynthesis, on a per unit leaf area basis, was higher in untranplanted trees than in transplanted trees. Photosynthetic gas exchange of red oak transplanted to the fill site, was below detectable levels. In contrast, one year after transplanting, photosynthesis, on a per unit leaf surface area, was greater in transplanted than in untransplanted chestnut oak (Table 3). However, whole plant photosynthesis in transplanted chestnut, shingle and English oak was less than in untransplanted plants because of greatly reduced canopy density: reduced number of leaves, reduced leaf size and lower rates of per unit leaf area net assimulation (Tables 1 and 3). Total plant photosynthesis is probably less two and three years after transplanting than for untransplanted shingle, English and red oak. Although, average leaf area was greater in transplanted than in untransplanted trees, net assimulation on a per unit leaf area basis was greatly reduced (Tables 1 and 3).

One year after transplanting, pre-dawn stem xylem water potential was less (more negative) in transplanted shingle and English oak trees than in untransplanted trees (Table 3). Chestnut oak responded differently; pre-dawn stem xylem water potential was higher in transplanted trees than in untransplanted trees. Pre-dawn stem xylem water potential in red oak transplanted into native soil had a pre-dawn xylem water potential similar to untransplanted red oak three years after transplanting. Red oak transplanted into a site with compacted clay subsoil had a more negative pre-dawn xylem water potential than untransplanted trees.

The establishment period is characterized by low canopy density. During establishment, the root system is being regenerated. The rate of development is species related. For instance, honeylocust regenerated more root mass 6 months after transplanting than English oak did 18 months after transplanting (6). Root regeneration in this study occurred under good conditions; the plants were heeled in a mixture of top soil and leaf mold. In a typical street tree planting site, the degree of root regeneration, and consequently the degree of shoot growth, would probably be significantly less due to poorer edaphic conditions.

Table 1. One and two year growth data from three oak species either transplanted or not. Untransplanted and transplanted trees were from the same nursery block, seed source, age and experienced similar cultural practices until transplanted. All trees were dug balled and burlapped and transplanted in April of 1992 (two years after transplanting) or in 1993 (one year after transplanting). Shoot length of transplanted plants was recorded each fall after transplanting. Shoot length of untransplanted plants was recorded for 1993 growing season only because lateral branches were dormant pruned as part of standard nursery tree training practices

Species	No. of plants	Years after transplanting	Average		Shoot length (cm)[3]		
			No. leaves[1]	leaf area (cm^2)[2]	1992	1993	Caliper
Quercus	5	0	28	36	20.0	-	5.4
imbricaria	8	1	13	5	5.6	-	2.3
	1	2	10	41	3.7	2.3	2.5
Q. robur	5	0	23	28	14.3	-	7.0
	3	1	10	10	1.8	-	2.7
	2[4]	2	9	58	10.5	2.7	3.5
Q. prinus	5	0	13	73	9.6	-	3.7
	5	1	9	32	4.6	-	2.0

[1] The average number of leaves on an individual terminal shoot of a lateral branch based on five shoots per tree.
[2] The average size of an individual leaf based on five leaves per tree.
[3] The average length of terminal shoots on five lateral branches per tree.

Table 2. Red oak growth data from transplanted and untransplanted trees. Trees were transplanted in spring 1990, data collected in summer 1993. Untransplanted and transplanted trees were from the same nursery block, seed source, age and experienced similar cultural practices until transplanted. All trees were dug balled and burlapped and transplanted in April 1991. Shoot length of transplanted plants was recorded each fall after transplanting. Shoot length of untransplanted plants was recorded for 1993 growing season only because lateral branches were dormant pruned as part of standard nursery tree training practices

Site	No. of	Years after	Average		Shoot length (cm)[4]			
conditions[1]	plants	transplanting	No. leaves[2]	leaf area (cm^3)[2]	1991	1992	1993	Caliper
Nursery	5	0	21	70	-	-	13.8	4.8
Native soil	5	3	21	91	8.6	3.7	12.5	4.0
Fill	5	3	10	49	2.5	3.5	4.7	2.6

[1] Site conditions were: Nursery site, Kokomo silty clay, Native, was disturbed site but in sod cover for last 15 years, Fill site, 2 to 3 meters of compacted clay subsoil. Trees were transplanted into this site within 6 months after fill was added.

[2] The average number of leaves on an individual terminal shoot of a lateral branch based on five shoots per tree.

[3] The average size of an individual leaf based on five leaves per tree.

[4] The average length of terminal shoots on five lateral branches per tree.

Table 3. Net photosynthesis and pre-dawn moisture potential of four oak species one and two years after transplanting

Species	No. of plants	Years after transplanting	Net photosynthesis[1] µmol $CO_2/m^2/s$	Pre-dawn stem xylem moisture potential (MPa)[2]
Quercus	5	0	.6.04	1.18
imbricaria	8	1	3.46	1.85
	1	2	2.41	-
Q. robur	5	0	10.38	1.38
	3	1	3.58	1.75
	2	2	3.19	-
Q. prinus	5	0	1.64	1.77
	5	1	3.28	1.56
Quercus	5	0	2.56	1.86
rubra	5	3	0.18	1.44

[1] Net photosynthesis is the average of two leaves per tree.

[2] Pre-dawn stem xylem moisture potential is the average of one terminal shoot of a lateral branch from each tree.

Ways to Increase Transplant Survival and Speed Establishment

There are many methods to improve transplant survival and reduce the establishment period; three methods will be briefly discussed.

Genetics

Foresters have begun developing root morphology grading systems to predict transplant survival (3,4). These systems are based on the number of permanent lateral roots at harvest; the greater the number, the greater the likelihood of survival. Certain mother trees have been identified that produce seedlings with higher than average number of lateral roots. Further, root morphology grading methods are quick and effective. However, less than 50% of the seedlings had five or more permanent lateral roots, the minimal acceptable number.

Container Production

Container-grown stock has higher transplant survival than bare-root stock because of higher root regeneration capacity (1). However, a problem with container-grown stock is root malformation. A malformed root system can lead to the development of girdling roots and mechanical instability. Producing plants in copper-treated containers is one method of controlling root malformation and maintaining high root regeneration potential. Copper-treated containers can be used to rapidly produce shade tree whips in containers (7). One possible benefit of container production of difficult-to-transplant species is the ability to profitably produce species previously unprofitable to produce. Typically these "unprofitable" species have coarse root systems that make them difficult to transplant and establish under nursery and urban landscape conditions. Coarse-rooted species are of interest to urban foresters for another reason; in general, coarse-rooted species are considered more drought resistant than fibrous-rooted species and thus well adapted to the harsh urban environment.

Coarse-rooted species can be grown successfully in the Ohio Production System (7). However, it is not known if these species can be established in the urban landscape. A cooperative project between The Ohio State University and several Ohio municipalities was begun in 1992 to determine transplanting and establishment success of coarse-rooted species in the urban landscape. Two year old container-grown planting stock produced under Ohio Production System conditions was transplanted in fall 1992 and spring 1993 to cooperating municipalities in Ohio. Initial results indicate that transplant success rate is high, but the establishment period, time before resumption of vigorous shoot growth, may be slow (Table 4). If successful, there is great potential for increasing the species diversity of the urban landscape.

Site Preparation

Two and three year results from shingle, English and red oak, indicate that site conditions have a significant effect on transplant survival and regrowth. Red oaks transplanted into native (undisturbed) soil have greater shoot extension, leaf size, trunk caliper and net photosynthesis than red oaks transplanted into an adjacent "fill" (disturbed) soil three years after transplanting (Tables 2 and 3). Further evidence of the importance of site quality on transplant success and regrowth comes from a small transplanting study conducted in 1991 with English and shingle oak (The two years after transplanting treatment in Tables 1 and 3). The English and shingle oaks were transplanted into a former gravel parking lot. The planting hole was dug with picks to get through six to eight inches of compacted gravel. The site was improved by amending the "native" soil backfill with an equal volume of composted municipal sewage sludge, 5 pounds elemental sulfur and five packets of slow release fertilizer per tree. Two years after transplanting, leaf size was greater in transplanted trees than in untransplanted trees. The take home message is that even poor quality sites can be easily and inexpensively improved.

Concluding Remarks

Transplanting research is difficult to conduct. Transplanting studies are typified by low numbers of experimental units and lack of adequate experimental controls; the author's examples described in this paper are not exceptions. The reason is not ignorance of proper experimental design, but rather the great (relative to most researcher's budgets) expense and complicated logistics involved in transplanting studies. Also, meaningful results are obtained three to five years after transplanting, rather than the first season after transplanting. This extended time frame inhibits university researchers, as tenure decisions for Assistant Professors are made before the five year results of the first transplanting studies are available for interpretation. Despite these challenges, tree establishment is one of the critical research issues for the urban landscape manager.

Literature Cited

1. Arnold, M. A. and D.K. Struve. 1989. Growing green ash and red oak in $CuCO_3$-treated containers increases root regeneration and shoot growth following transplant. J. Amer. Soc. Hort. Sci. 114:402-406.
2. Johnson, P. S., S.L. Novinger and W.G. Mares. 1984. Root, shoot and leaf area growth potentials of northern red oak planting stock. For. Sci. 30:1017-1026.
3. Kormanik, P.P. 1986. Lateral root morphology as an expression of sweetgum seedling quality. For. Sci. 32:595-604.

Table 4. Growth of plants produced under Ohio Production System conditions after transplanting into urban landscape sites in Ohio

Species	Location	Number of plants	Number dead	Height (cm)	
				Initial	1993
Gymnocladus dioica	Cleveland(1)	9	0	152	127
	Powell	9	0		187
	Upper Arlington	8	0		263
Koelreuteria paniculata	Powell	6	0	160	188
	Upper Arlington	9	0		222
Nyssa sylvatica	Cleveland(1)	5	2	205	168
	Cleveland (2)	4	0		218
Quercus alba	Powell	8	0	125	156
	Upper Arlington	7	0		153
Q. bicolor	Cleveland(1)	10	0	224	244
	Powell	10	1		142
	Upper Arlington	3	0		263
Q. coccinea	Powell	2	0	182	215
	Upper Arlington	2	0		272
Q. imbricaria	Powell	1	0	200	215
	Upper Arlington	2	0		258
Q. macrocarpa	Cleveland(1)	6	0	137	150
	Cleveland(2)	30	0		206
	Powell	1	0		225
	Upper Arlington	3	0		245
Q. michauxii	Powell	8	0	231	276
	Upper Arlington	1	0		310
Q. prinus	Upper Arlington	1	0	197	260
Q. robur	Powell	4	0	194	238
	Upper Arlington	2	0		263
Q. rubra	Powell	5	0	220	242
	Upper Arlington	1	0		235
	Columbus	2	0		315
Q. shumardi	Cleveland(1)	6	0	185	209
Q. velutina	Cleveland(1)	8	0	183	178
	Powell	5	0		172
	Upper Arlington	3	0		286
Taxiodum distichum	Cleveland(1)	12	0	174	180
	Cleveland (2)	6	0		200
	Powell	10	0		182
	Upper Arlington	6	0		199

4. Kormanik, P.P., J.L. Ruehle and H.D. Muse. 1988. Frequency distributions of seedlings by first order lateral roots: a phenotypic or genotypic expression. In Proc. of the 31st Northeastern Forest Tree Improvement Conference and the 6th Northcentral Tree Improvement Assoc. July 7-8; University Park, Pa. University Park, PA: School of Forest Resources, Pennsylvania State University. pp 181-187.
5. Struve, D. K. and B.C. Moser. 1984. Root system and root regeneration characteristics of pin and scarlet oak. HortScience. 19:123-125.
6. Struve, D.K., T.D. Sydnor and R. Rideout. 1989. Root system configuration affects transplantability of honeylocust and English oak. J. Arboric. 15:129-134.
7. Struve, D.K. and W.T. Rhodus. 1990. Turning copper into gold. Amer. Nurseryman. 172(4):114-135.
8. Struve, D.K. and R.J. Joly. 1992. Transplanted red oak seedlings mediate transplant shock by reducing leaf surface area and altering carbon allocation. Can. J. For. Res. 22:1441-1448.
9. Sutton, R.F. 1990. Root growth capacity in coniferous forest trees. HortScience 25:259-266.
10. Watson, G. W. and T.D. Sydnor. 1987. The effect of root pruning on the root systems of nursery trees. J. Arboric. 13:126-130.

Advantages of the Low Profile Container

Daniel C. Milbocker

The low profile container was developed to eliminate the problems of poor survival of field dug trees, circling roots and the thin trunks of conventional container grown trees. These problems were reduced or eliminated and other benefits were realized. Because of the greater container diameter, circling roots and tree tip-over were reduced in nurseries and tip-over was reduced and survival enhanced in landscapes.

Large scale commercial culture of nursery stock in containers is relatively recent, beginning with fruit cans less than forty years ago (4). Container culture has progressed to the familiar plastic container of today. Large trees are not as widely grown in containers as other nursery stock because of their aggressive roots. These roots either circle within the container or escape to the soil beneath them (1). Removal of escaped roots often destroys the plant. Circling roots have the potential for girdling the trunk in later years and eventually killing the tree (4). Consequently, most large tree production remains in field soil even though a portion of these trees fail to survive transplanting. Survivability has been improved by briefly growing field-grown plants in containers before marketing but the risk of survival is shifted to the producer and the extra operation is expensive. Tree production in containers needs improvement.

Elimination of transplanting from the field requires culturing of trees in containers from liners to landscape sized trees (5). Thin crooked trunks, circling roots and the extra cost of lengthy growing in containers discourages this practice, but overproduction of more easily grown nursery stock has renewed an interest in this type of production. The low profile container was developed to solve the typical problems of growing large nursery stock in containers. It provides an efficient means of growing large nursery stock suitable for landscaping.

Trees naturally grow with as much as 90% of their roots in the top twelve inches (30 cm) of soil which results in a lateral spread much greater than its depth (3). Conventional containers have depths approximately equal to their diameter and for large containers this is deeper than most of the natural root system. A container with a shallow depth and a compensating larger width more nearly fits a natural root system. Such containers are called low profile containers (2). These containers have two types of advantages, those resulting during production and those resulting during transplanting into the landscape.

Daniel C. Milbocker is with the Hampton Roads Agricultural Experiment Station, Virginia Polytechnic Institute and State University, Virginia Beach, VA 23455.

Setting Up the Low Profile Container

Any well drained location with a supply of irrigation water can be used for growing plants in low profile containers. When a field is used, rows approximately three feet (0.9 m) in width are tilled, leveled and covered with a 36 inch (91 cm) wide strip of black polyethylene. Six mil thickness is necessary to prevent puncturing during planting and penetration by roots. The entire area can also be covered with polyethylene as is practical for smaller containers but 6 mil thickness must be used. A band of 8 to 10 mil black polyethylene is placed on a form to hold it in a circular position and inverted on the plastic covering. Forms for growing trees are 30 inches (76 cm) in diameter and hold 25 gal (95 liters) of medium though smaller or larger sizes are possible. The growing medium should not contain sand or field soil. The excess weight will break the roots when grown plants are lifted by the trunk. Pine bark and sphagnum peat 3:1 have most commonly been used though peat has been replaced by other forms of humus with success. These media have been amended with 1.5 lb (0.68 Kg) Micromax (Grace Sierra, Milpitas, CA) and 5 lbs (2.3 Kg) of ground limestone per cubic yard (0.76 m3) as is recommended for smaller containers. Optimum fertilization for a 25 gal (95 liter) container has been 0.44 lb (200 g) 18-6-12 Osmocote per season. A tree is planted in the container at filling and the form is withdrawn leaving a filled and planted container. The next container is spaced at least 6 inches (15 cm) from its neighbor to allow air pruning of roots. Trees can be irrigated by overhead sprinklers or two spray type emitters. A fast growing tree will completely permeate the growing medium within one year and most slow growing trees within 2 years. After roots have completely permeated the growing medium, the tree and its container can be moved as a marketable unit. Overwatering may delay root permeation, particularly at the beginning and end of the growing season when water use is reduced by dormancy, reduction of foliage and cooler weather.

Production Advantages

Time

One of the greatest advantages of container culture is that growing conditions can be optimized. Nutrients and water are controlled so that trees grow to marketable sizes in a minimum of time and can be more nearly grown to order along with other nursery stock. Low profile containers provide the soil volume necessary for fast growth.

Tip Over

Trees are top heavy. When grown in conventional containers, the force required to tip them over easily occurs during windy weather. The greater diameter of low profile containers increases the force required for tip-over so that trees must be very large and winds must exceed 50 mph.

Circling Roots

If given adequate room, roots seldom circle. When they meet an obstruction, they branch. The low profile container provides adequate space for lateral root growth. Air pruning at the lower edge encourages branching and growth of a large number of small roots that completely permeate the growing medium. Circling seldom occurs.

Thick Trunks

The trunk diameter depends on the number of roots and branches on the tree. Highly branched roots produce the vascular system necessary for thick trunks. If branch pruning is not excessive, trunks are more like those of field grown trees. Trunks are strong enough to remain straight and become crooked only if allowed to remain leaning after a wind. Correction of leaning trees is required of container or field grown trees and is not specific to low profile containers.

The low profile container enables the nurseryman to grow a large tree in a shorter time than with other methods. It eliminates circling roots and thin trunks characteristic of container grown plants. A tree or any other large nursery stock can now be container grown without the previous weaknesses associated with container grown plants.

Transplanting Advantages

Trees grown in low profile containers are easier to transplant than field-grown or conventional container-grown trees.

Lighter to Handle

Trees grown in pine bark media are lighter in weight than field-dug trees even though all the root system comes with container-grown trees. Lightness provides more flexibility in handling, and less time and equipment are required for moving trees. Most trees are light enough for manual handling if the root mass is not saturated with water. A two inch caliper tree can be picked up by the trunk with the root mass balancing the top weight.

Planting Holes are Easier to Dig

Shallow holes are easier to dig than deep holes. When transplanting a tree grown in a low profile container, the hole should be dug no deeper than the container height. Rapid transplanting recovery depends upon leaving the soil below the tree undisturbed, which provides an immediate capillary water supply to the tree.

Easy Recovery from Transplanting

Low, profile container-grown trees have three to four times more bottom area than field-dug or conventual container-grown plants. The amount of capillary water available to the tree is proportional to its area of contact with undisturbed soil beneath the plant. Trees transplanted from low profile containers have successfully survived when transplanted in full leaf without supplemental irrigation.

Self Supporting

Low profile, container-grown trees planted on undisturbed soil resist tip-over because of their wider base. Trees have withstood 50 mph winds immediately after transplanting without additional support.

Recovery is Natural

Roots enlarge after transplanting due to normal growth. Circling roots of conventional, container-grown trees have been observed to grow into a solid mass of wood of the size of the original container with roots growing at random from their ends. Roots of low profile container grown trees continue to grow radially and slightly downward from the root mass. A few roots become dominant and enlarge to become a naturally appearing root system.

Summary

The low profile container offers a convenient and easy means of growing trees with sturdy trunks and without circling roots. The benefits of growing trees in low profile containers extend to landscapers. Trees are easier to plant, recover sooner and develop a better root system.

Literature Cited

1. Harris, R. W., W. B. Davis, N. W. Stice and D. Long. 1971. Root pruning improves nursery tree quality. J. Amer. Soc. Hort. Sci. 96:105-108.
2. Milbocker, D. C. 1991. Low profile container for nursery-grown trees. HortScience 26:261-263.
3. Pearson, R. W. 1971. Significance of rooting pattern to crop production and some problems of root research. pp. 247-270. In Carson, E. W. (ed.) The Plant Root and Its Environment. University Press of Virginia, Charlottesville, Virginia. 691 pp.
4. Whitcomb, C. E. 1984. Plant Production in Containers, Ch. 3. Lacebrook Publications Rt. 5, Box 174, Stillwater, Oklahoma 638 pp.

Elimination of Circling Tree Roots During Nursery Production

Bonnie Lee Appleton

When trees are grown in containers during nursery production, roots often begin to circle on the outside of the root ball against slick and/or smooth container walls. If not mechanically disrupted when the trees are transplanted, circling roots may enlarge to the point of stressing or killing trees by girdling. Nursery containers treated with a copper compound, or containing a variety of wall modifications, are a means to reduce or prevent root circling during production.

Girdling roots—roots that grow around tree stems and other roots—may shorten a tree's life span by constricting the vascular system and restricting water and nutrient movement, and by failing to adequately anchor trees (14,17,29). Girdling roots may start as roots that circle in structurally restrictive planting holes or planting holes with glazed clay walls, or as new lateral roots that develop behind the ends of primary roots cut during field-grown nursery stock harvesting (28). They also may start as roots that circle on the outside of the root ball for trees grown in containers with slick and/or smooth walls.

A common planting recommendation relative to container-grown trees is mechanical disruption of the root ball by slicing through or cutting away any circling roots found when the container is removed (11,14). The value of these practices is questionable, with limited and contradictory research conducted primarily using shrubs (6,26,31). To insure greater tree transplant success, it therefore seems appropriate to reduce or eliminate circling root formation during nursery production.

In-ground Production Alternatives

Several in-ground alternatives to conventional field production of bareroot and B&B (balled-in-burlap) trees have been developed, including pot-in-pot, in-ground plastic containers, and in-ground fabric containers. Each of these methods can influence directional root development.

Pot-in-pot

The new pot-in-pot system (22) involves sinking an outer or sleeve pot into the ground, and inserting a second pot, the actual production pot that is harvested with the tree, within and resting upon the lip of the sleeve pot. The production container often has vertical basal ribs, or may be copper-treated (see below), to reduce root circling.

Bonnie Lee Appleton is an Extension Nursery Specialist with the Hampton Roads Agricultural Research and Extension Center, Virginia Tech, 1444 Diamond Springs Road, Virginia Beach, VA 23455.

In-ground Plastic Containers

Traditionally, trees could not be grown in the ground in single plastic containers due to drainage problems. A rigid plastic container has been developed with rows of small holes around the container sides, and throughout the container bottom, to minimize drainage problems. This container is new. No comparative tests have thus far been reported, but the potential for circling root formation appears minimal.

In-ground Fabric Containers

In-ground fabric containers, or grow bags, are the oldest of several hybrid field/container production options (21). Numerous comparative studies of these containers vs. conventional field or container production have been conducted. Some tree species respond better to in-ground fabric container production, while others respond poorly (12,13,15,16,18).

Though in-ground fabric containers usually prevent circling root formation, circling roots have been observed at the bottom of these containers (personal observation). To address this issue, alternative fabrics and container designs have been developed. One new container is composed of fabric with holes of a size designed to allow only small roots to penetrate for absorption of water and nutrients from the surrounding soil, but not to impede harvesting. The potential for circling root formation should be minimal with this fabric change.

Above-ground Production Alternatives

Modified Container Designs

A variety of approaches, including container wall ribs, holes, baffles and other root deflecting or pruning devices have been developed to modify conventional straight, smooth-walled, rigid plastic containers to reduce or eliminate circling root formation.

The various wall modifications, and a flexible poly bag container, have significantly reduced circling root formation on many species of plants (1,2,27,29,30), although sometimes with conflicting results relative to shoot growth (19,29). Once planted to the landscape, the effectiveness of the modifications in enhancing new root generation has been found to be species specific (27).

A porous-walled container with pin-hole perforations randomly punctuating the container walls produced roots superior to those in nonporous smooth and nonporous ridged containers (20). Air-root pruning behind the perforations prevented circling root formation except where the plastic was denser and container air porosity was limited.

Soil Sock Containers

A new above ground container, that combines wire baskets used to protect field-grown tree root balls with a porous foam-rubber liner, is called the "Soil Sock" (25). While the liner insulates the roots against temperature extremes, it allows air penetration, thereby air-pruning the roots and preventing circling root formation. The container sits above ground for production, but is reported by the manufacturer to be entirely plantable, and is currently being tested by the author for transplantability and circling root reduction.

Copper-coated Containers

One final strategy for the reduction or elimination of circling root formation is the use of rigid plastic containers with copper-coated interior walls (23). Applied to the walls in a carrier, the copper is absorbed by the root tips from the carrier. The copper acts as a growth regulator, inhibiting root tip growth and stimulating branching. The manufacturer claims that root tips are not killed by the copper as they are with air-pruning.

The effectiveness of the copper has been demonstrated on a large number of trees and shrubs (3,4,5,7,8,9,10,23,24). Results range from virtually no visible roots on the outside of root balls, to roots whose tendency to circle is stopped after one to two inches of growth. No impairment of root growth out into the surrounding soil has been reported for trees and shrubs after copper-coated container removal and field transplanting.

Sources of Nursery Production Alternatives

In-ground fabric containers

Lacebark Inc., PO Box 2383, Stillwater, OK 74076;
(405) 377-3539

Root Control Inc., 7505 N. Broadway, Oklahoma City, OK 83116;
(405) 848-2302

In-ground plastic containers

Rootmaker Grounder - Lacebark

Pot-in-Pot

Lerio Corp., Mobile, AL
(800) 457-8112

Nursery Supplies, 250 Canal Road, Fairless Hills, PA 19030
(215) 736-3641

Modified containers

Lerio, Nursery Supplies, Menne, others

Soil Sock Container

BetterBilt Products, PO Box 559, Addison, IL 60101
(800) 544-4550

Thomas' Nursery, Rt. 2, Box 180A, Enterprise, MS 39330
(601) 659-9259

Copper coating

Spin Out™ - Griffin Corp., PO Box 1847, Valdosta, GA 31603-1847
(800) 237-1854

Literature Cited

1. Anonymous. 1991. Container designed to aid plant's root development. Nursery Manager 7:22, 24.
2. Appleton, B.L. 1989. Evaluation of nursery container designs for minimization or prevention of root circling. J. Environ. Hort. 7:59-61.
3. Arnold, M.A. 1992. Timing, acclimation period, and cupric hydroxide concentration alter growth responses of the Ohio production system. J. Environ. Hort. 10(2):114-117.
4. Arnold, M.A. and E. Young. 1991. CuCO3-painted container and root pruning affect apple and green ash root growth and cytokinin levels. HortScience 26(3):242-244.
5. Beeson, R.C. and R. Newton. 1992. Shoot and root responses of eighteen southeastern

woody landscape species grown in cupric hydroxide-treated containers. J. Environ. Hort. 10(4):214-217.

6. Blessing, S.C. and M.N. Dana. 1987. Post-transplant root system expansion in *Juniperus chinensis* L. as influenced by production system, mechanical root disruption and soil type. J. Environ. Hort. 5:155-158.
7. Case, G.N. and M.A. Arnold. 1992. Cupric hydroxide-treated containers decrease pot-binding of five species of vigorously rooted greenhouse crops. Proc. SNA Research Conf. 37:94-98
8. Flanagan, P.C. and W.T. Witte. 1991. Effects of chemical root pruning on root regeneration and cellular structure of viburnum root tips. Proc. SNA Research Conf. 36:46-49.
9. Flanagan, P.C. and W.T. Witte. 1991. Development of a growth curve model for root regeneration of chemically root pruned viburnum. Proc. SNA Research Conf. 36:99-101.
10. Flanagan, P.C. and W.T. Witte. 1992. Root regeneration of sawtooth oak after chemical root pruning. HortScience 27:1166-1167. (Abstract)
11. Flemer, W. 1982. Successful transplanting is easy. Amer. Nurseryman 145:43-55.
12. Fuller,D.L. and W.A. Meadows. 1987. Root and top growth response of five woody ornamental species to fabric Field -Grow™ containers, bed height and trickle irrigation. Proc. SNA Res. Conf. 32:148-153.
13. Fuller, D.L. and W.A. Meadows. 1988. Influence of production systems on root regeneration following transplanting of five woody ornamental species. Proc. SNA Research Conf. 33:120-125.
14. Gouin, F.R. 1984. Updating landscape specifications. J. Environ. Hort. 2:98-101.
15. Harris, J.R. and E.F. Gilman. 1991. Production method affects growth and root regeneration of Leyland cypress, laurel oak and slash pine. J. Arboric. 17:64-69.
16. Harris, J.R. and E.F. Gilman. 1993. Production method affects growth and post-transplant establishment of 'East Palatka' holly. J. Amer. Soc. Hort. Sci. 188:194-200.
17. Holmes, F.W. 1984. Effects on maples of prolonged exposure by artificial girdling roots. J. Arboric. 10:40-44.
18. Ingram, D.L., U. Yadav, and C.A. Neal. 1987. Production system comparisons for selected woody plants in Florida. HortScience 22:1285-1287.
19. Neuman, S. and W. Follet. 1987. Effects of container design on growth of Quercus laurifolia Michx. Miss. Agr. Expt. Sta. Bul. 12:7.
20. Privett, D.W. and R.L. Hummel. 1992. Root and shoot growth of 'Coral Beauty' cotoneaster and Leyland cypress produced in porous and nonporous containers. J. Environ. Hort. 10:133-136.
21. Reiger, R. and C.E. Whitcomb. 1982. A system for improving root development and ease of digging field grown trees. Okla. Ag. Expt. Sta. Res. Report P-829:10-13.
22. Roberts, D.R. 1993. How pot-in-pot systems save time, money. Nursery Manager 9(6):46, 48, 50.
23. Struve, D.K. and T. Rhodus. 1990. Turning copper into gold. Amer. Nurs. 172:114-125.
24. Svenson, S.E. and D.L. Johnston. 1992. Faster growth of *Radermachera sinica* after transplanting from copper-treated liners. Proc. SNA Research Conf. 37:100-101.
25. Tilt, K. 1992. Growing solutions. Amer. Nurs. 176:62-65.
26. Wade, G.L. and G.E. Smith. 1985. Effect of root disturbance on establishment of container grown *Ilex crenata* 'Compacta' in the landscape. Proc. SNA Research Conf. 30:110-111.
27. Warren, S.L. and F.A. Blazich. 1991. Influence of container design on root circling, top growth, and post-transplant root growth of selected landscape species. J. Environ. Hort. 9:141-144.
28. Watson, G.W., S. Clark, and K. Johnson. 1990. Formation of girdling roots. J. Arboric.

16:197-202.
29. Whitcomb, C.E. 1984. Container design: Problems and progress, pp 107-130. In Plant production in containers. Lacebark Publications, Stillwater, OK.
30. Whitcomb, C.E., and J.D. Williams. 1985. Stair-step container for improved root growth. HortScience 20:66-67.
31. Wright, R.D. and D.C. Milbocker. 1978. The influence of container media and transplanting technique on the establishment of container grown *Rhododendron* cv. 'Hershey Red' in landscape plantings. SNA Research Conf. 5(2):1-7.

Post-transplant Root Growth and Water Relations of *Thuja Occidentalis* from Field and Containers

Michael N. Dana and Steven C. Blessing

Post-transplant root growth and shoot water relations of *Thuja occidentalis* were studied. Shrubs were either transplanted from the field as balled and burlapped or from 7.6 liter (2 gallon) plastic containers. Container-grown plant roots were either left undisturbed or mechanically disrupted by vertical slashing or "butterflying." Root expansion was increased by mechanical disruption. However, post-transplant roots of container plants, whether mechanically disrupted or not, were less uniformly distributed around the root ball than those of balled and burlapped plants. Water stress was most severe on balled and burlapped plants immediately after transplanting, compared to the container plants. The maximum stress level was reduced significantly by an irrigated establishment period preceding the imposition of moisture deficit. Mechanical disruption had no significant effect on moisture stress levels of container plants.

Rapid root regeneration is important for the survival of transplanted woody plants. This is especially true in the urban landscape where environmental stress factors may reach severe proportions. Maintenance, especially irrigation, may be infrequent or totally lacking. Under such conditions, rapid root regeneration to effect maximum soil exploration may be critical to plant survival.

Woody nursery stock has historically been field grown and handled bare root or balled and burlapped. More recently, container grown plants have become important because they permit production and handling of nursery stock faster and more economically (6).

When a plant is moved balled and burlapped, the majority of its root system may remain in the field. In studies of several shade tree species, less than 5% of tree root systems were retained by the digging process (18). The entire root system of a container grown plant, however, remains intact and essentially undisturbed. Thus, logic suggests that container grown plants should offer better likelihood of transplant success. However, it has been observed container-grown plants often establish poorly when moved to the landscape (5,10). This may be due to the assumed tendency for roots of container-

Michael N. Dana and Steven C. Blessing are with the Department of Horticulture, 1165 Horticulture Bldg., Purdue University, West Lafayette, IN 47907-1165.

grown plants to continue to grow in a circle after transplanting, only slowly expanding radially into the soil (8,13).

Other factors may contribute to poor transplantability of container-grown plants. Container-grown plants are produced in a manner designed to maximize size of the plant shoot and leaf system including both frequent watering and fertilization. The use of light, soilless mixes in containers creates a media interface discontinuity when a plant is transplanted into heavier soil such that root extension into the surrounding soil may be inhibited (6,16).

Several cultural practices have been developed to address some of these problems. Use of soil amendments to improve the structure and aeration of backfill for transplants is common although several studies have shown no consistent improvement in plant re-establishment and growth from the use of amendments (4,12). In clay soil, scratching or cutting of the planting hole wall may help prevent glazing (3). Mechanical disruption of the root ball is recommended to prevent circling roots and encourage rapid root development radially (8,9). Two common types of mechanical disruption are vertical slashing, and "butterflying" in which the bottom half of the root ball is cut through vertically and the root system physically pulled apart while being placed into the planting hole(10).

In our earlier work with *Juniperus chinensis* (2) we found that after 8 weeks following transplanting in loam soil, balled and burlapped plants and slashed container plants had similar new-root dry weight, while undisrupted container plants had significantly less new-root dry weight, but greater new-shoot extension growth. After 16 weeks, balled and burlapped plants still exhibited the greatest new-root dry weight in loam soil, with both container plant types showing significantly less new-root dry weight. In clay soil, the relationship was the same, but between the container types, undisrupted containers had developed significantly more new root dry weight than the slashed container root system.

These results were complicated by problems with the experimental material and by the mid-June planting date which missed the spring season when most woody plants exhibit a flush of new shoot, then root growth. During this study, we did observe an apparent tendency for container grown plants to produce root systems that were one-sided or only rooted in limited sections of the root ball. Balled and burlapped plants, on the other hand, seemed to produce a much more uniform, three-dimensional root system. Regeneration of a three-dimensionally uniform root system is a likely key to successful transplant establishment. The amount of soil explored by a plant's root system is directly related to the uniformity of distribution and extent of root growth.

The objectives of part one of this study were 1) to repeat our earlier work with a different species, 2) to focus on early post-transplant root regeneration, 3) to assess the uniformity of new root distribution of container grown and balled and burlapped plants in different soil types and, 4) to evaluate butterflying as a mechanical disruption technique.

Water deficits (14,15) are the primary stress associated with transplanting, and one of the major aspects of transplant shock (5). There are, however, little data available relating transplant root type, or mechanical disruption of container plants to water relations.

The objectives of part two of this study were 1) to evaluate the effect of production system on post-transplant water relations; 2) to evaluate the effect of mechanical root ball disruption on the water relations of newly transplanted container grown plants, and 3) to determine whether a well-irrigated establishment period prior to drought stress was beneficial for plant water status compared to no establishment period prior to imposition of drought.

Materials and Methods

Part One: Experiment 1-1

Seven liter (2 gal) container grown plants and similar sized balled and burlapped plants of *Thuja occidentalis* 'Little Gem' were obtained from Gaar's Nursery in Chesterfield, Indiana. The container-grown plants were grown in a medium consisting of 1/3 field soil and 2/3 pine bark. The balled and burlapped plants were field grown in a Brookston silty-loam soil. Average rootball size for the container grown plants was 18 cm wide and 15 cm deep. Average diameter of the balled and burlapped rootballs was 20 cm. The balled and burlapped plants were within industry standards for rootball size and were selected based on their similarity to the container-grown plants in top size. Balled and burlapped root ball size was as nearly identical to container-grown root ball dimensions as could be practically achieved. Both container-grown and balled and burlapped plants were from common stock at the nursery and produced in a typical industry production system.

Treatments were a factorial set consisting of four root conditions and two soil types. The root conditions were balled and burlapped (B&B), container-grown with an undisturbed rootball (CG), container grown with mechanical disruption (slashing) to the rootball (S/CG) and container-grown with a "butterflied" root system (B/CG). Mechanical disruption consisted of 6 vertical slashes 2.5 cm (1 in) in depth evenly spaced around the rootball (Fig. 1). "Butterflying" consisted of a vertical cut with a spade through the bottom half of a rootball followed by gently pulling the rootball open at the cut while placing it into the planting hole (Fig. 2). "The butterflied" root system was placed over a small ridge of soil in the planting hole in order to assure as much soil to root ball contact as possible, and to keep the halves apart. The soil types were a Fox loam and a Brookston silty clay, both in West Lafayette, Indiana.

Planting sites were prepared using a Howard rotovator and 24.4 kg/100 square meters (50 lbs/1000 square feet) of 8-24-24 was pre-plant incorporated. Planting holes were hand dug and approximately 2x the size of the rootballs in diameter. No soil amendment or fertilizer was added to the backfill.

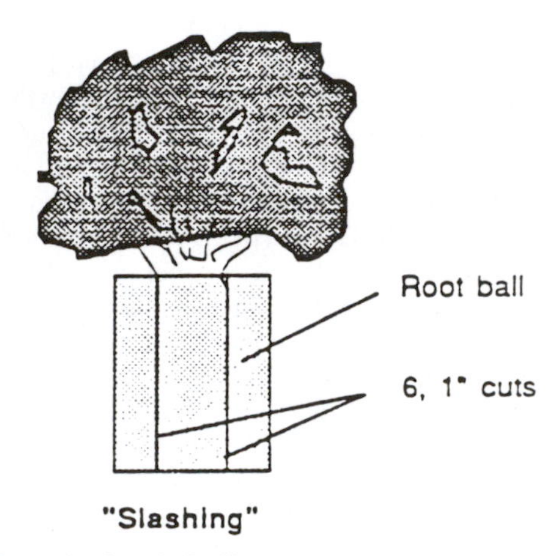

Figure 1 — Example of a "slashed" root system.

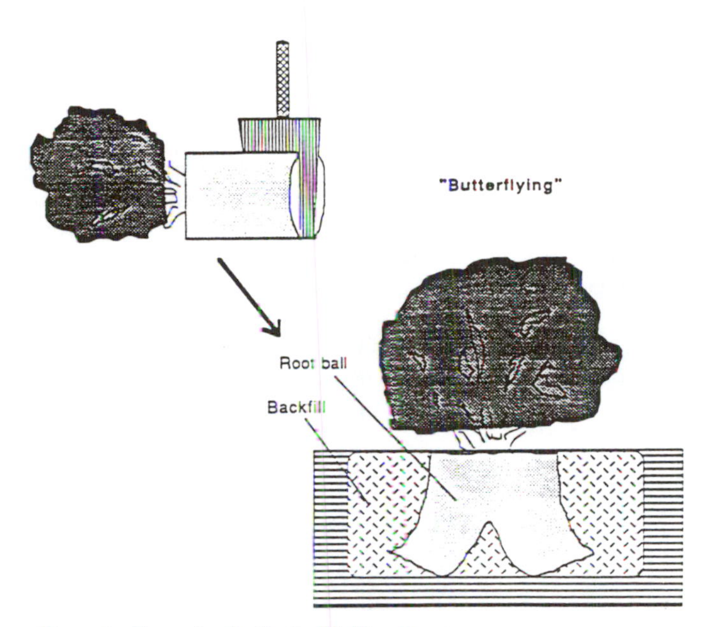

Figure 2 – Example of a "butterflied" root system.

Plants were arranged in 4 rows of 9 plants with 1.5 m (5 ft) between rows and 1 m (3 ft) between plants in the row. Each set consisted of 9 replicates of four root conditions. Root conditions were completely random within each soil type. Planting took place in mid-May.

The planting site was watered to field capacity at planting time and natural rainfall was monitored and supplemented as needed to a minimum level of 2.5 cm (1 in) water/per week throughout the experiment. Weeds were controlled by hand. No disease or insect control was used.

Both sets of plants were harvested 40 days after planting. Plants in the Fox loam soil were harvested using a backhoe with a 67 cm (30 in) bucket. The plants in the Brookston silty clay soil were harvested using a different technique, necessitated by the physical charateristics of the soil. Circular trenches were dug surrounding each plant immediately outside of the root zone. These trenches were filled with water continuously over a 12 hour period in order to thoroughly saturate the soil area surrounding the root balls. Excess soil was then washed away from each root ball using a gentle stream of water before removing the plants from the field for data collection.

For data collection, root ball surfaces were divided into 9 sectors using specially constructed wire baskets. Sectors were randomly assigned to the surface of each rootball with the following restrictions. Sector 9 was always assigned to the bottom of the rootball. The remainder of the rootball surface area was divided into 8 equal sectors. Sectors 1-4 were sequentially assigned in a counter-clockwise fashion to the top half of the rootball. Sectors 5-8 represented an equal area on the bottom half of the rootball (excluding the area represented by sector 9) and were sequentially assigned such that sector 5 was directly below sector 1, sector 6 was directly below sector 2, etc. (Fig. 3). Sector 9 represented the the circular bottom of the rootball for the CG and S/CG treatments. For B/CG sector 9 also included the inside area of the butterflied cut. For B&B, sector 9 was defined as the surface area below an imaginary plane dissecting the rootball 5 cm (2 in) above the base of the rootball (Fig. 4). For a 20 cm (8 in) sphere (the approximate size of the balled and burlapped rootballs) this area is proportionally equal to the amount of area represented by the circular bottom of a CG rootball.

The total number of growing white root tips in each sector was counted. Ignoring the longest root, the next three longest roots were measured and a mean computed for each sector. All new root growth was then shaved at the original rootball surface, washed and dried at 70°C to obtain new root dry weight for each sector.

Data for each set of plants were analyzed separately as a completely random design. Means separation was done by Duncan's multiple range test for treatments within each soil type. A split-plot, one-way analysis of variance was used to test differences between soil types. Differences between sector combinations within individual treatments were tested using single degree of freedom contrasts procedures and an appropriate F-test.

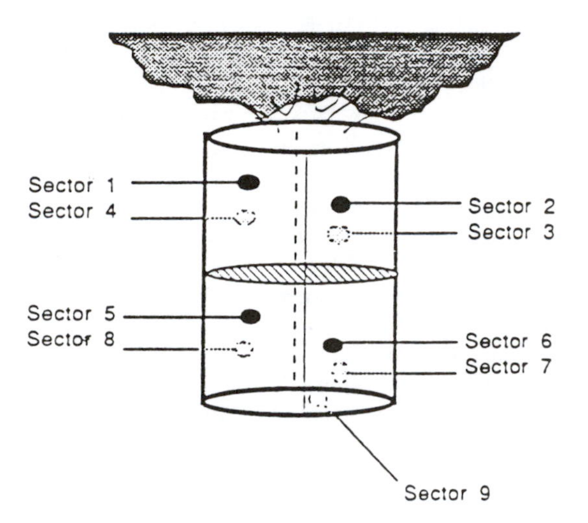

Figure 3 – Sector assignment for a container grown root ball.

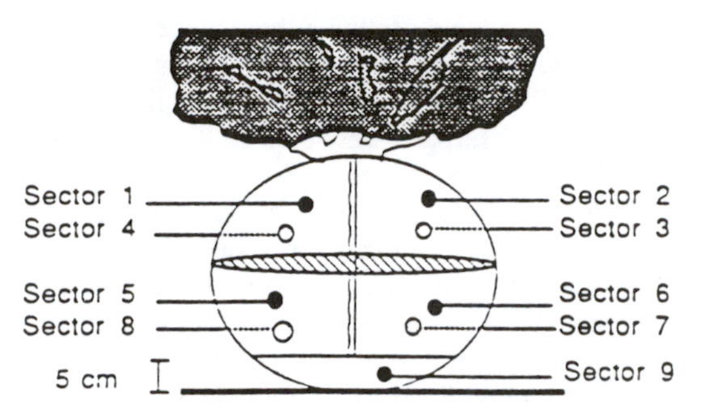

Figure 4 – Sector assignment for a balled and burlapped (field grown) root system.

Variability within treatment sectors was evaluated via a mean separation (Duncan's) of transformed (log10 (x+1)) variance values for the individual replicates (17).

Part Two: Experiment 2-1

Shoot water potential was measured at regular, pre-determined intervals on the plants in experiment 1-1. A stem and foliage sample was excised from each of three randomly selected replicate plants for each treatment and soil type. Water potential was measured using a Scholander pressure bomb. Pre-dawn readings were taken on all sampling dates.

Experiment 2-2

Seven liter (2 gal.) CG plants of *Thuja occidentalis* 'Little Gem' and similar sized B&B plants were obtained from Gaar's Nursery at Chesterfield, Indiana in mid-May. Plants were selected to be as similar in top and rootball size as could be practically achieved. A raised bed measuring 10 m (32 ft) long, 3 m (10 ft) wide and 0.6 m (24 in) deep was constructed in an open field at West Lafayette, Indiana. The bed was filled with a sandy Fox loam soil. The soil had been previously prepared using a Howard roto-vator and 24.4 kg/100 square meters (50 lbs/1000 square feet) of 8-24-24 had been pre-plant incorporated.

Planting techniques and treatments were the same as in Part One. A set of 36 plants was planted in the raised bed on May 22. Plants were arranged in 3 rows of 12 plants each with 0.75 meters (2 1/2 ft) between plants and rows. Treatments were completely randomized within the bed.

Experiment 2-3

A second raised bed was constructed as described above at the same location in late June. A second set of plants was obtained and planted in the raised bed on July 3. Planting techniques and treatments were the same as for experiment 2-2.

Experiments 2-2 and 2-3 varied as follows. Both beds were watered thoroughly after planting to hasten the settling of the soil and to assure that no air pockets had formed around the root balls. Experiment 2-2 was watered by over-head irrigation as necessary between May 22 and July 2 to supplement natural rainfall to a level of 2.5 cm (1 inch) water/per week. On July 3rd (the planting date for experiment 2-3) both beds received a thorough watering. Beginning July 4, both experiments were subjected to a simulated drought. Beds were not irrigated and natural rainfall was excluded by draping a sheet of white opaque plastic over the beds during threatening weather.

During the experiments shoot water potential was measured at regular intervals on plants in both raised beds with a Scholander pressure bomb. A stem and foliage sample was excised from each of three randomly selected replicate plants of each treatment. Pre-dawn readings were taken on all sampling dates for both beds. Additional readings at approximately 11 am, 3 pm and 7 pm (whole day readings) were also taken on day 1, 8, 46 and 65 for experiment 2-2 and days 5, 23 and 49 for experiment 2-3.

Soil moisture was also measured at regular intervals for both raised beds.Three samples were taken at a depth of 8-15 cm (3-6 in) from each raised bed. Samples were placed in a plastic bag, transported to the laboratory where a field weight was recorded for each sample. The samples were oven dried at 100°C and re-weighed to determine percent soil moisture. The experiments were terminated on August 22, which was day 91 for experiment 2-2 and day 49 for experiment 2-3. Analysis of variance procedures, with mean separations by Duncan's multiple range test were applied to both water potential and soil moisture data sets.

Results and Discussion

Part One

The data for three root growth parameters in two soil types are presented in Table 1. After 40 days of growth the largest number of roots and greatest root dry weights were produced by B/CG in clay soil. B/CG produced approximately 75% higher root dry weights and root counts than CG in clay soil. S/CG also seemed to follow the same trend of more root growth than CG in clay soil although the differences were not statistically significant. The same trends were present in loam soil, but not as distinct. Root length did not vary among treatments which indicates that root elongation was not inhibited by the greater quantity of roots elongating or being initiated in the B/CG and S/CG treatments.

B&B root growth was generally less than S/CG or B/CG and similar to CG in both soil types. Although differences were not always significant, this trend contrasts with our results in *Juniperus* (2). A possible explanation is that although the *Thuja* root balls were within American Association of Nurserymen (AAN) standards, the size of the plants was considerably smaller than what would be normally dug as a balled and burlapped plant. This may also be an indication that there is considerable variation in adaptability to a particular production system even among species considered easy to transplant. These results also question the validity of the recommended root ball size given by the AAN. More specific recommendations for different species may be appropriate.

Root disruption clearly stimulated root regeneration in this experiment. Although the apparent differences between CG and B/CG or S/CG were not always statistically significant, greater replication may have confirmed the trends in treatment effects. The B/CG treatment increased root growth over S/CG primarily by increasing the root ball surface area exposed to the soil.

The B&B and B/CG root systems had different surface areas than the CG or S/CG plants. This was compensated for by expressing root data on the basis of the surface area of the original root ball placed in the ground at transplanting. There were no differences between B/CG and S/CG in root dry weight or root count on an area basis (Table 2). Root dry weight of B/CG was significantly higher than CG on an area basis in clay soil. This indicates that butterfying stimulated greater root growth in addition to increasing

Table 1. Root growth in clay and loam soil for all sectors combined

	Loam			Clay		
Root	Root Count	Root Dry Wt.	Root Length	Root Count	Root Dry Wt.	Root Length
Condition	(no.)	(grams)	(cm.)	(no.)	(grams)	(cm.)
B&B	77.6 a[x]	111.8 a	12.2 a	99.8 a	236.5 a	17.2 a
CG	76.2 a	139.0 ab	20.7 b	131.8 ab	285.9 a	27.3 b
S/CG	109.0 a	194.2 b	20.0 b	228.0 c	542.5 b	29.6 b

[x]Mean separation within columns using Duncan's multiple range test ($P<.05$).

Means based on 9 replications per root condition for loam soil, 8 replications per root condition for clay soil.

the root ball surface area. S/CG also produced numerically greater root growth on an area basis although differences were not always statistically significant.

Table 3 presents comparisons between soil types for root count, root dry weight and root length data expressed per 100 sq cm. All root conditions produced greater total root dry weight in clay. CG, S/CG and B/CG also produced higher root counts and CG and S/CG had greater root lengths in clay than in loam soil. B/CG was the only root condition to produce significantly more root dry weight for sector 9 (sector data not presented) in clay soil. S/CG produced significantly higher root counts and root dry weights in clay soil for all sectors except 9.

These data indicate that *T. occidentalis* is well adapted to clay soil regardless of planting technique or production system. *T. occidentalis* is known to adapt well to alkaline soils, and is native to wetland habitats (7). Clay soils in Indiana tend to simulate these conditions since they tend to hold water and have an alkaline pH reaction. It is also significant that root length was not inhibited by clay soil for any treatment. The significant increase in root length of CG and S/CG in clay soil is in direct contrast to what we observed with *Juniperus* (2). It would appear that species adaptability to the soil envi-

Table 2. Root growth per 100 square cm. of root ball surface in clay and loam soil for all sectors combined

	Loam		Clay	
Root Condition	Root Count (no.)	Root Dry Wt. (grams)	Root Count (no.)	Root Dry Wt. (grams)
B&B	6.2 a[x]	8.9 a	7.9 a	18.8 a
CG	6.9 ab	12.6 ab	12.0 ab	25.9 ab
S/CG	9.9 b	17.6 b	15.2 bc	37.4 b
B/CG	8.1 ab	15.8 b	16.5 c	39.4 b

[x]Mean separation within columns using Duncan's multiple range test (P<.05).
Means based on 9 replications per root condition for loam soil, 8 replications per root condition for clay soil.

Table 3. Root growth per 100 square cm. of root ball surface[x] for all sectors combined in different soil types

	Root Count (no.)		Root Dry Weight (grams)		Root Length (cm.)	
Root Condition	Loam	Clay	Loam	Clay	Loam	Clay
B&B	6.2	7.9 NS[y]	8.9	18.8*	12.2	17.2 NS
CG	6.9	12.0*	12.6	25.9*	20.7	27.3*
S/CG	9.9	15.2*	17.6	37.4*	20.0	29.6*
B/CG	8.1	16.5*	15.8	39.4 NS	23.1	29.6 NS

[x]Root length is not on an area basis.
[y]Means based on 9 replications in loam soil and 8 replications in clay soil.
A one-way analysis of variance is non-significant (NS) or significant at P<.05(*) for means within root condition.

ronment may be more important than the differences in soil characteristics. However, root extension in this and the previous study was through soil that had been previously cultivated, and was kept moist. Root extension into an undisturbed soil can not necessarily be predicted based on the data presented here.

Uniformity of new root distribution was examined by a log tranformation of variances by sectors, and by single degree-of-freedom contrasts. The variance testing focuses on lateral distribution while the contrasts highlight depth variability.

The variance data (1) offer no evidence that root disruption increases the uniformity of root growth of container plants as was hypothesized. On the contrary, root disruption tended to increase variability.

Contrast comparisons of sector combination totals for root dry weight in loam soil are presented in Table 4. There were no significant differences in either soil type for B&B (1). CG and S/CG produced more root dry weight in loam soil for sectors 5-8 than 1-4. Sector 9 had significantly less root dry weight (on an area basis) than other sector combinations for B/CG. CG followed the same trend although differences were not always significant. S/CG produced significantly greater root dry weight in sector 9 than 1-4.

The comparisons using contrasts indicate that all treatments produced the majority of their root growth in clay soil on the bottom half of the root ball (sectors 5-8). In loam soil the same was true of B&B, CG and S/CG. Root length was uniform throughout CG and B/CG in both soil types. S/CG produced significantly greater root length in loam soil for sector 9, while B&B produced significantly greater lengths in sector 9 for both soil types. These data support the trends in root count and root dry weight observed for sector 9.

Part Two

Figs. 5 and 6 present the pre-dawn water potential readings for experiment 2-1 (taken on plants in experiment 1-1). Shoot water potential readings in both soil types were

Table 4. Single degree of freedom contrasts of root dry weight per 100 square cm. of root ball surface area for different sector combinations in loam soil

Root Condition	Sector Comparisons[y] upper vs lower	upper vs bottom	lower vs bottom	upper & lower vs bottom
B&B	4.80[x]	4.80	11.03	7.92
	11.03	11.18	11.18	11.18
CG	9.16	9.16	18.82	13.99
	18.82**	8.20	8.20**	8.20
S/CG	9.51	9.51	17.68	13.60
	17.68**	18.08**	18.08	18.08
B/CG	19.58	19.58	24.92	22.49
	24.92	5.82**	5.82**	5.82**

[x]Means based upon 9 replications per root condition. Single degree of freedom contrast is significant at P<.05(*), or at P<.01(**).

[y]Upper = sectors 1 thru 4; lower = sectors 5 thru 8; upper and lower = sectors 1 thru 8; bottom = sector 9.

remarkably uniform throughout the data collection period. In clay soil, water potential readings fluctuated between -2.9 and -6.9 bars over the period (Fig. 6). Significant differences among treatments occurred on only one sampling date (May 28). In loam soil, readings fluctuated between -2.8 and -5.2 bars for all treatments (Fig. 5). Significant differences between treatments occurred on three sampling dates (May 17, 26 and 28). B&B water potentials were significantly more negative than at least one of the other treatments on each of these dates. On May 17, B/CG was significantly lower than CG or S/CG in loam soil. This confirmed the trend of greater water stress in butterflied plants which had been evident on the two previous sampling dates, but which was not statistically significant at the 5% level. On May 28, B&B water potentials were significantly lower than other treatments in both soil types. CG were the least stressed on the same day. After the first two weeks, there were no discernable treatment effects.

These data provide no evidence that any of these four root conditions of nursery stock are subject to excessive water stress after transplanting when frequent watering procedures are followed. Typical wetting/drying cycles were observed in both soil types. B&B treatment tended to be under more water stress during the first two weeks than the other treatments in loam soil (Fig. 5). This trend may have been partially due to the slightly larger shoot sizes of the B&B plants which would have been subject to more transpirational loss than the other treatments. In clay soil, the B&B plants were under significantly more drought stress only on May 28. This indicates that in spite of the severe reduction of roots associated with moving a plant balled and burlapped, it is not subject to excessive stress if adequate water is supplied. Given this observation it it not surprising that neither S/CG, or B/CG were under more water stress than CG. The amount of root loss associated with slashing or butterflying is apparently insignificant if water is readily available.

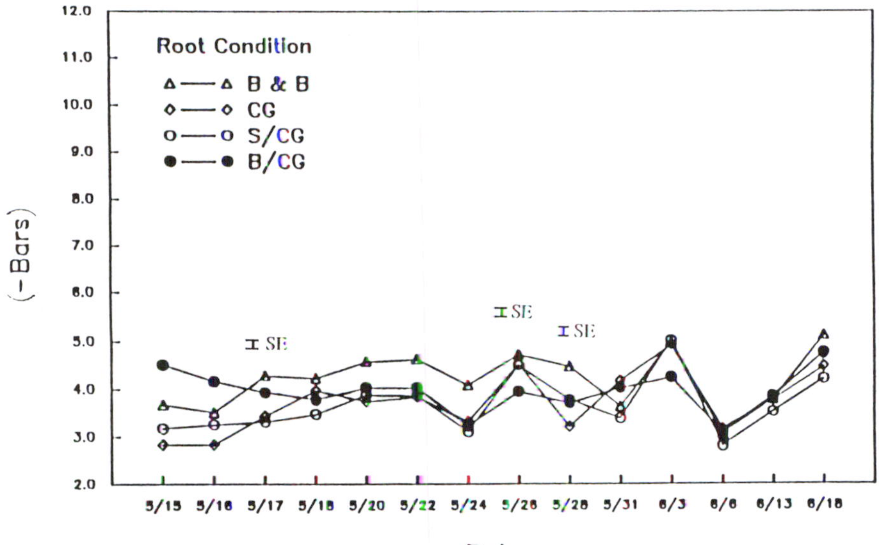

Figure 5 — Post-transplant pre-dawn water potentials of *Thuja* in loam soil.

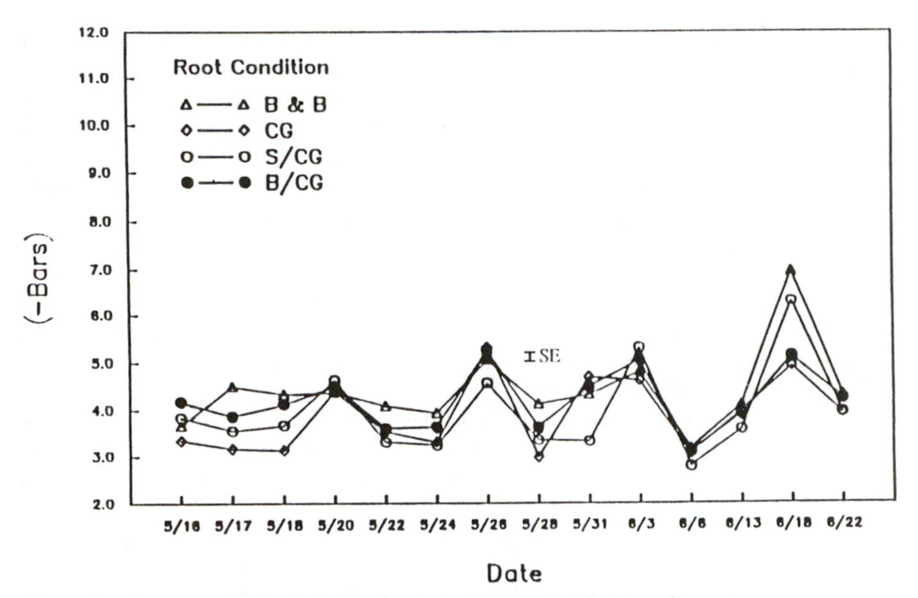

Figure 6 — Post-transplant pre-dawn water potentials of *Thuja* in clay soil.

It is interesting to observe that no water potential readings approached 0 bars, even though soil moisture was often at or near field capacity. Pre-dawn readings taken the morning following planting ranged between -2.7 and -4.5 bars (Fig. 5) even though soil moisture was near field capacity. Throughout these experiments water potentials did not rise above -2.2 bars after any drying cycle. Water potentials of -2 or -3 bars have been reported to be detrimental to plant growth in some species (14). It would appear that even under good moisture-availability conditions, the transplanting process imposes an important level of water stress on all nursery stock, regardless of root condition.

Experiments 2-2 and 2-3 were designed to test plant responses to drought following either a 6 week irrigated establishment period (exp. 2-2), or drought immediately following transplanting (exp. 2-3).

Pre-dawn water potential readings for experiment 2-2 (Figure 7) showed significant treatment differences on every sampling date. B&B had lower shoot water potential than at least one of the other treatments on every sampling date during the first two weeks. There were no significant differences among CG, S/CG and B/CG during the first two weeks except on May 24 when S/CG had a higher water potential. Through the last 6 weeks of the experiment no obvious trends developed among the treatments, although B&B tended to have the most negative readings. The beginning of simulated drought on July 4 did not cause any consistent treatment differences. Water potentials did not begin to decrease until two weeks into the drought period. After 6 weeks of simulated drought, water potentials for all treatments were in the range of -7.0 to -7.5 bar. At termination of the experiment on August 22, there were no visible signs of drought injury on the plants.

Experiment 2-3 with immediate post-transplant drought, resulted in a two-week lag period for a substantial decrease in water potentials to become evident (Figure 8). Even then, there were no differences among treatments. Only after 4 weeks of drought did treatment effects become evident and then it was the B&B plants which exhibited water

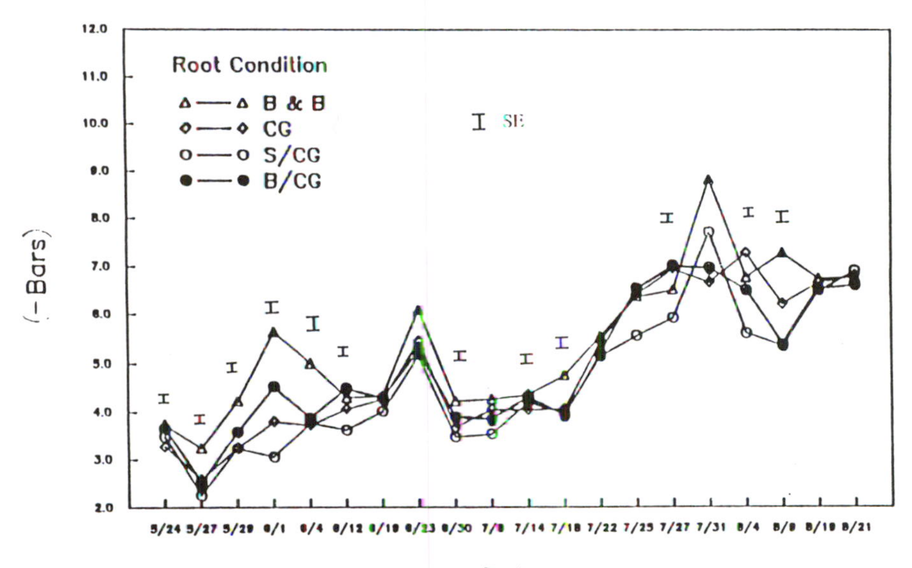

Figure 7 — Post-transplant pre-dawn water potentials of *Thuja* subjected to a delayed simulated drought.

potential readings in the range of -10 to -12 bars. No differences were noted among container treatments.

In experiment 2-2 drought was not imposed until 6 weeks after planting. The 6-week period allowed enough time for significant root regeneration to begin (1) and also simulated the natural high rainfall period a spring-planted shrub might encounter before a summer drought. During the establishment period, a characteristic wetting/drying cycle was observed. The B&B treatment tended to develop greater water deficits during the first two weeks as was the case in experiment 1-1. The data from experiment 2-2 may have produced more significant differences due to the more uniform soil environment provided by the raised bed. No significant patterns developed among CG, S/CG or B/CG during the establishment period. Rainfall exclusion was begun on July 4. Water potentials did not begin to decline until after two weeks of drought (Fig. 7). The fact that water potentials remained fairly stable after August 1 is not surprising since soil moisture remained relatively stable (data not presented).

The data from the whole-day water potential readings (1) support many of the trends apparent in the pre-dawn readings. However, one notable finding was that the B/CG treatment reached significantly ($P< .05$) higher water deficits during mid-day or early evening than S/CG or CG even though there were no significant differences in the pre-dawn readings. The fact that there were no differences among these treatments at the pre-dawn sampling for their respective dates, indicates that the plants were able to recover by the following morning. These episodes of increased stress could slow transplant establishment by stomatal closure resulting in a temporary reduction of photosynthesis.

The plants in experiment 2-3 were planted on July 3 and subjected to drought immediately after an initial watering. As observed in experiment 2-2, there was a 2-week lag

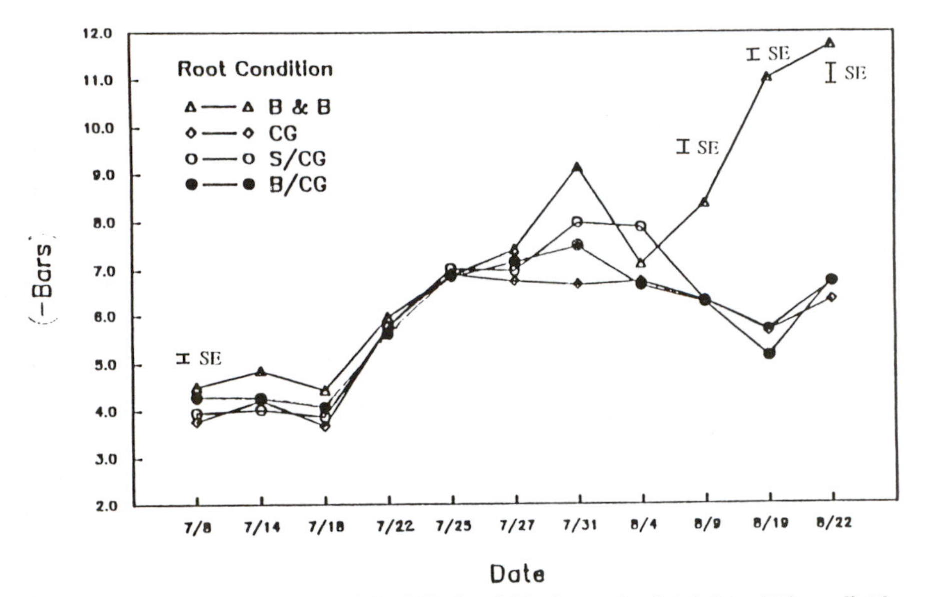

Figure 8 — Pre-dawn water potentials of *Thuja* subjected to a simulated drought immediately post-transplanting.

period before water potential readings began to decline. This would seem to indicate that the initial development of water stress in the treatments was not altered by a previous establishment period. Apparently the amount of root growth produced after 6 weeks was not sufficient to delay the onset of water stress. On July 27, pre-dawn water potential readings in both raised beds were between -6.0 and -7.5 for all treatments. After August 1, water potential readings for the three container grown treatments remained stable. The rapid rise in water deficits of the B&B treatment after August 4 suggests that an irrigated establishment period is critical to the survival of balled and burlapped stock.

Several aspects of these results were surprising. *T. occidentalis* is considered a drought sensitive species by most authorities (7). Yet, in both these experiments water deficits were very slow to develop. There was no evidence of roots expanding outside of the raised bed, and soil moisture was uniform at all depths (1). *T. occidentalis* is known to adapt to dry, barren habitats within some parts of its range, but these colonies are believed to be true ecotypes rather than characteristic of the species (11). Other researchers have reported severe water stress in container plants subject to a drought after only a few days when planted in a sandy loam (5). After 6 weeks of simulated drought in our Fox loam there were no visible signs of drought injury except for the B&B treatment in experiment 2-3. These observations suggest that the permanent wilting point for Thuja occidentalis is between -7.0 and -11.5 bars.

Root system data were not collected in these experiments because of the complicating factor of significant shoot removal for sampling. However, it was observed that all treatments in experiment 2-2 produced considerable new root growth. This contrasted with experiment 2-3, in which all container treatments produced noticeable new root growth, while few or no new roots developed on the B&B root balls. Apparently the lag time between drought induction and development of substantial water stress was sufficient to allow the commencement of regeneration of a substantial root system.

Conclusions

In this work on *Thuja occidentalis*, root disruption was beneficial in terms of increasing the quantity of new root growth in both soil types compared to either undisrupted containers or B&B stock. Uniformity of root distribution among container plants, however, was not increased by root disruption. All treatments produced the majority of root growth on the bottom 1/2 of the root ball in both soil types. This observation may be a reflection of the degree of "pot-boundness" of the container-grown treatments since roots tend to congregate toward the bottom of a pot-bound container-grown plant. Root disruption appears to derive its benefits from stimulation of root growth in areas where active root tips are already present. Butterflying may have an extra benefit since the root ball is physically oriented such that the majority of its growing root tips are brought closer to the surface where soil environmental conditions may be favorable.

The B&B treatment produced relatively uniformly distributed root growth, but generally less quantity of roots than other treatments. All treatments adapted well to clay soil, which may indicate that species adaptability rather than cultural technique or production system is of primary importance in post-transplant root development.

Severity of moisture stress in container grown *Thuja occidentalis* was not affected by root ball disruption in any consistent way. A two-week period elapsed between imposition of drought conditions and measurable increase in stress on the plants, whether they had experienced an establishment period or not. This may be useful knowledge for scheduling minimal irrigation for survival of transplants. The decision to disrupt container-grown plants at transplanting should be made based on a desire to enhance root regeneration, or from fear of girdling roots, without concern that the method will result in extra water stress.

Literature Cited

1. Blessing, Stephen C. 1987. Characterization of first season post-transplant root regeneration of balled and burlapped and container grown nursery stock. M.S. Thesis, Purdue University, West Lafayette, Indiana, USA.
2. Blessing, Stephen C. and Michael N. Dana 1987. Post-transplant root system expansion in *Juniperus chinensis* L. as influenced by production system, mechanical root disruption and soil type. J. Environ. Hort. 5(6):155-158.
3. Clark, J. R. 1981. Planting and aftercare of trees and shrubs - an update. Plants and the Landscape. Purdue University Cooperative Extension Service. 4(3):1-4.
4. Corley, W. L. 1984. Soil amendments at planting. J. Environ. Hort. 2(1):27-30.
5. Costello, L. and J. L. Paul. 1975. Moisture relations in transplanted container plants. HortScience 10:371-372.
6. Davidson, H. and R. Mecklenberg. 1981. Nursery Management. Prentice Hall, Inc., Englewood Cliffs, N.J.
7. Dirr, M. A. 1977. Manual of Woody Landscape Plants. Stipes Publishing Co., Champaign, Illinois.
8. Flemer, W. 1982. Successful transplanting is easy. American Nurseryman. 145:43-55.
9. Gouin, F. R. 1983. Girdling by roots and ropes. J. Environ. Hort. 1(2):48-50.
10. Gouin, F. R. 1984. Updating landscape specifications. J. Environ. Hort. 2(3):98-101.

11. Habeck, J. R. 1958. White-cedar ecotypes in Wisconsin. Ecology 39:457-463.
12. Hummel, R. L. and C. R. Johnson. 1985. Amended backfills: their cost and effect on transplant growth and survival. J. Environ. Hort. 3(2):76-79.
13. Ingram, D. L. and H. van de Werken. 1978. Effects of container media and backfill composition on the establishement of container-grown plants in the landscape. HortScience 13:583-584.
14. Kramer, P. J. 1983. Water Relations of Plants. Academic Press, NY.
15. Levitt, J. 1980. Responses of Plants to Environmental Stresses. 2nd ed., Vol. 2. Academic Press, N. Y.
16. Mecklenberg, R. A. 1983. An overview of problems. J. Environ. Hort. 1(1):26-27.
17. Steele, R. G. D. and J. H. Torrie. 1960. Principles and Procedures of Statistics. McGraw-Hill, Inc., NY, NY.
18. Watson, G. W. and E. B. Himelick. 1983. Root regeneration of shade trees following transplanting. J. Environ. Hort. 1(2):52-54.

PART II

MANAGING ROOT SYSTEMS AND SOIL ENVIRONMENTS OF ESTABLISHED TREES

Urban Soils: An Overview and Their Future

Phillip J. Craul

Urban soils are those existing or designed soils in the urban environment exhibiting anthropeic characteristics that may present poor plant conditions, and challenge our imperfect knowledge and expertise for the amelioration of these conditions to meet the requirements of adequate to optimum plant (mainly tree) growth.

There are several conditions in urban soils that continue to present problems to landscape architects, urban foresters, arboriculturists, horticulturists, arborists and landscape contractors. These conditions include: the loss or the lack of natural structure with a decrease in favorable plant root growth conditions including crusting, decreased aeration and drainage and waterholding capacity; interrupted organic matter and nutrient cycling; a high degree of variability in properties and their characteristics unless designed and installed as a homogenous mass; elevated reaction (pH); presence of anthropeic materials and contaminants; and highly modified soil temperature regimes. We will discuss several of these in more detail to provide clarity and allow presentation of possible solutions with our present knowledge and technology.

Major Problems

Urban soils have been disturbed, manipulated or handled in some way that changes or modifies their properties and attendant characteristics; there are very few urban soils that do not exhibit evidence of modification in some degree (1).

Loss of Structure Resulting from Handling or Construction/ Restoration Activity Leading to Several Modifications

Structure is the physical property of aggregation of the primary particles of sand, silt and clay into units called peds that renders form or a matrix to the soil. It determines (along with texture to a lesser degree) the total pore space, the pore size distribution and hence the proportion of air-filled (macropores) and water-filled pore (micropores) space under field conditions, which in turn, influences the drainage of the soil, the aeration status (oxygen diffusion rate), infiltration, waterholding capacity, and the ease of root penetration. These modifications are discussed so that solutions may be better applied.

Compaction. This process is the compression of the porous, unsaturated soil. The

---

Phillip J. Craul is Professor of Soil Science, SUNY-College of Environmental Science and Forestry, Syracuse, NY 13210 and Visiting Professor of Landscape Architecture, The Graduate School of Design, Harvard University, Cambridge, MA 02138.

units of structure, the peds, are at least partially crushed. Total pore space is reduced, accompanied by a reduction in the proportion of macropore space and usually an increase in the proportion of micropore space. Bulk density increases as a result. Stockpiling and respreading of topsoil has a similar effect on the structure.

Decreased aeration. The proportional shift in volume from macropores to micropores tends to reduce the gaseous diffusion pathways since the micropores are generally water-filled for most of the moisture content range. Their higher degree of discontinuity as compared to the macropores also reduces the diffusion pathways (2,4). Thus, the oxygen diffusion rate is reduced in compacted soils, even if only the surface is compacted.

Decreased drainage. Most water flow in the soil occurs under saturated or near saturated conditions in the macropores. These are destroyed in varying degrees by compaction or by the handling of soil. The degree of destruction is dependent on the percentage of water stable aggregates in the soil and by the moisture content when disturbed or handled. Since water must flow through smaller pores in the compacted soil, the drainage becomes more restricted. Infiltration is affected in much the same way by compaction as is internal drainage.

Decreased waterholding capacity. Any significant decrease in total pore space will decrease waterholding capacity. There are situations under relatively light to moderate compaction where the increase in the proportion of micropore space may increase waterholding capacity; these situations are rare in urban areas.

Decreased root penetration. With the loss of macropores and perhaps the larger micropores, the roots have greater difficulty in penetrating the soil. Roots can penetrate only pores that have the same or larger diameters as the root tips. Loose soils present little resistance to root penetration.

Interrupted Organic Matter and Nutrient Cycling

The urban environment does not usually provide for the same cycling of seasonal organic material that occurs in the forest, either because the soil surface is mostly covered with inorganic, impervious materials or the litter is removed as a nuisance. Thus, the soil is not provided with the organic material that decomposes into humus , which enhances the development of structure, acts as source of energy for soil organisms, helps buffer the soil, and provides nutrients on decomposition. Additionally, the urban soil profile may not be connected to a natural parent material or bedrock from which nutrients may also be released on mineral weathering. In some cases the urban soil "parent material" may be of questionable origin for sustaining plant growth. We must be sensitive to the need for organic matter amendments to urban soil and possibly fertilization because of these conditions. However, the latter sometimes receives too much attention from the tree/green industry in light of Harris' findings (3).

Spatial and Vertical Variability

The constructional history of most urban sites is sufficiently complex and unknown so that the conventional means of making a soil survey are not adequate (1). Conventional soil inventory and survey depends on the attributes of parent material, topography, vegetation and some knowledge of the underlying hydrology. Observation and measurement of these attributes provide some degree of predictability of certain soil characteristics and their distributional patterns. Boundaries can be delineated thus differentiating the soil characteristics. Several of the mapping attributes are lacking or hidden in the urban environment, revealed only with intensive, expensive sampling and analysis, unless detailed engineering and survey records are available. The visible or measurable attributes are so complex that no patterns are discernible. The complexity of soil conditions are most times ignored with one standard planting design applied indiscriminately over the landscape. This problem presents more challenge to the landscape architect, arboriculturist, and contractor than any other in urban soil.

Lack of Dissemination of Existing Technology

We presently have much knowledge of urban soils at our disposal which is mal-distributed. Thus, some of the solutions and their applications suggested by appropriate scientists are not widely known outside of ourselves; the people who need the information most are not fully exposed to it. This situation has not changed very much since a 1983 meeting at Longwood Gardens, Pennsylvania, of the Soils Working Group of the now defunct Pinchot Environmental Forestry Consortium. As scientists, this is partially our responsibility. But publication in scientific journals is not enough. We need to present information at conferences and symposia like today at Morton Arboretum, but also on a more local level. There needs to be more information produced for technical manuals and videos. The International Society of Arboriculture has initiated this type of activity with success.

The Future Urban Soil Situation

There are changing conditions within and outside of our usual areas of scientific interest that will influence what we do with our urban soils in the future. These involve doing a better job with the information we now have; the influence on our work by the concept of sustainable design from the landscape architect's viewpoint or from the ecologist's viewpoint of environmental impact; and, the fruits of more sharply defined research accomplished with decreasingly available and more restrictive funds.

Improved Soil Specifications and Design

Soil specifications are commonly described in terms of what many people conceive as the old standby of "topsoil." For example, a current project engineer and architect both requested the project soil be "of natural loam topsoil, well drained with appropriate organic matter content, of good tilth, and containing little debris, stones and roots." If we examine this specification more closely, it is totally inadequate from the soil and arboricultural aspects. Our present soil and arboricultural knowledge requires greater detail. The following items in Table 1 comprise an adequate list contained in a more valid soil specification. We have to agree that engineering and design specifications are far more complex and detailed than they were years ago. It must be understood that the soil specification has evolved in detail as well — and yet people still use the same specification as that of the turn of the century!

Particle size distribution. The specification of a USDA textural class such as sandy loam or loam is too broad for modern day applications, especially for the sandy textures. The sand particle diameter ranges from 2 mm to 0.05 mm; a difference of two magnitudes. The various soil characteristics of waterholding capacity, infiltration rate and bulk density can potentially differ significantly from a texture of very coarse loamy sand through the sandy loam classes, or the lower left segment of the conventional texture triangle (Fig. 1). Therefore, the specification should include the percentage of soil passing the various sieve sizes. This assures the proper proportion of the coarse versus fine sands, including the silt and clay so that the soil has the proper characteristics without being self-compacting. Those soils with a relatively high proportion of fine and/or very fine sand

Table 1. Urban soil specifications

TEXTURE VERSUS PARTICLE SIZE DISTRIBUTION
- Rather than simply loam, etc., give percentage passing the sieve sizes.
- Provide with a particle size distribution curve.

ORGANIC MATTER CONTENT
- Content important, but source of the organic matter amendment more important
- The difficulty in the "selling" of composted sewage sludge.

SOIL REACTION
- Care because of the danger from high pH.

WATER-STABLE AGGREGATES
- Not a common or usual item in soil specifications.
- Gives a measure of the soil resistance to structure loss in handling. Proven as an important property in the maintenance of structure in agricultural soils.
- Many laboratories do not know how to do the analysis and do not have the specialized equipment for analysis.
- Education required of landscape architects, engineers and contractors.

SOLUBLE SALT CONTENT
- 600 ppm usually used as a caution level; danger at 1000 ppm.

STONINESS AND DEBRIS CONTENT
- Should be limited by specific percentage and size limits.

CONTAMINANT CONTENT
- Generally, there should be no pesticide residues, and a low level of heavy metals.

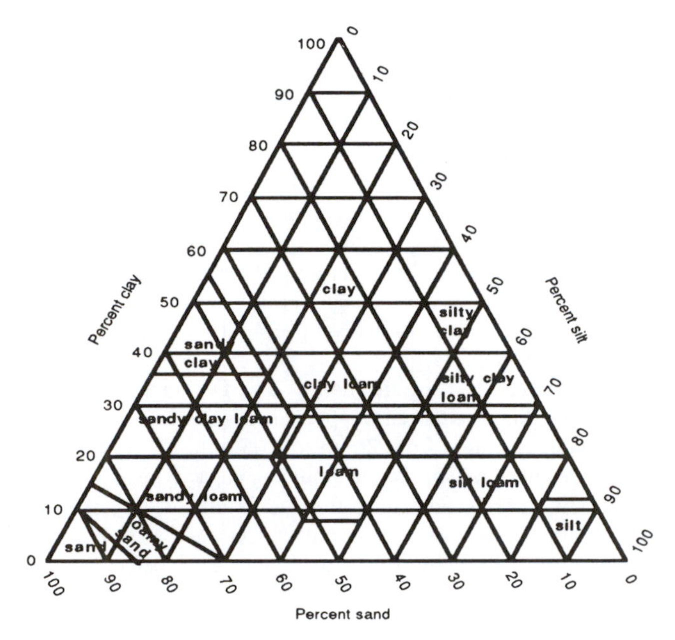

Figure 1 — The USDA texture triangle

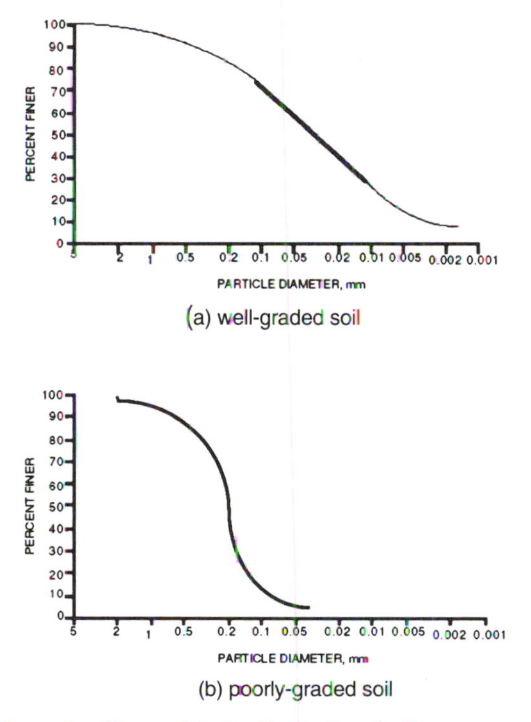

(a) well-graded soil

(b) poorly-graded soil

Figure 2 — The particle size distribution of soil.

and/or silt causes the soil to have a high self-compaction potential. The specification could reference to the ASTM D-422-63 standard analysis to assure conformance.

Fig. 2 illustrates the contrast in particle size distribution between a well-graded and a poorly graded soil. For compaction resistance the poorly-graded soil is preferred, but being sandy it has the disadvantages of low waterholding capacity and low or moderate fertility. The well-graded soil may have favorable waterholding capacity and fertility, but may be compaction-susceptible. However, a word of caution is required. A fine or very fine sandy loam soil may have the same shape of curve as in 2a. This soil could be relatively infertile with a high bulk density. On the other hand, a well-aggregated light silt loam or loam may have a similar curve.

Organic Matter Content. Organic matter being an important component in all soils, it is usually included in a soil specification. Most forest soils have about 4 to 5 percent organic matter content in the topsoil and some agricultural soils may have as much as 10 to 15 percent. Topsoil is usually specified in order to provide for sufficient organic matter. However, with the decrease in the availability of topsoil in many localities, it is necessary for organic matter amendment. Shredded or milled peat moss has been a popular amendment, but this product has become relatively expensive, and substitutes are now sought. These substitutes will be discussed later.

Soil Reaction (pH). This soil property is given in the specification because of its overall indication of the general chemical condition of the soil and also to make certain that extreme levels of pH do not exist. These levels usually indicate potential problems of nutrient deficiencies or toxic conditions. The range is normally given as pH 6.5 to 7.5.

Water-stable Aggregates. This characteristic as a soil specification may seem new or strange to many of you. But, soil scientists, agronomists and agricultural engineers have known the importance of water-stable aggregates to productivity for many years (1,4). Soil aggregates are the units of structure we see on the spade when digging in the soil. They range in size from one or two inches or more across to very fine ones only several millimeters in diameter. The larger ones are usually broken into smaller aggregates without difficulty whereas some of the smaller ones resist fragmentation even when wetted. These latter are water stable-aggregates cemented by organic mineral complexes and resist the force of raindrop impact and the effects of tillage; thus, they maintain the favorable structure important to productive soils (2). These aggregates, though small, will persist through topsoil stripping, stockpiling and respreading in most cases, and are important in reducing the compaction potential of a manipulated soil. Unfortunately, most laboratories do not analyze soil for water-stable aggregates nor have the specialized equipment to do so. However, it is necessary to encourage the use of this valuable tool in our soil diagnosis. Empirical data on water-stable aggregates need to be developed for urban soils.

Soluble Salt Content. In most humid regions soluble salt content does not pose a problem to root growth except where de-icing salts are used. In semi-arid and arid regions, the accumulation of soluble salts is sufficient to cause a low osmotic potential in the soil, thus causing difficulty to the root for absorption of moisture and nutrients. Some authorities use 600 ppm soluble salts as a caution level and 1000 ppm or more as a dangerous level for most unadapted plants. Some locations in northern metropolitan areas will have soluble salt levels greater than the latter value.

Stoniness and Debris Content. Stones and other solid debris interfere with root elongation, dilute the soil and increase the difficulty of cultivation. Because debris is commonly present in large volumes in urban soils, almost all specifications should require the screening of soil materials.

Contaminant Content. Soil contaminants include all pesticides, heavy metals, phytotoxic compounds, and anthropeic materials that are present in concentrations above certain threshold levels. Generally, heavy metals are of general concern only if food stuffs are to grown in the soil, but if present in excess amounts, even at low concentrations for some compounds or ions, root growth is adversely affected. Copper, lead, and zinc are examples along with ammonia and boron as common examples of toxic substances. Pesticide residues can create problems, especially the more persistent herbicides. Insecticides and most fungicides are not usually a problem to plants but may of concern where human activity is to occur on the site. There is danger of contaminant-laden dust being ingested, absorbed through the skin or enter the lungs.

In some cases the specifications for a project may go beyond the soil itself to assure a consistent attention to those details necessary for a successful conclusion to a design. An example is the preliminary specifications of the 30-year restoration plan for the Harvard Yard in Cambridge, Massachusetts. Here, specifications are given for both the on-site soil design and tree stock treatment in the nursery. These are summarized in Table 2. Many of the details still are in the development stage, but the information gives an indication of the scope of the specifications for this rather comprehensive, and perhaps unique, project.

More Sustainable Designs with Minimal Off-site Impact

The usual procedure to import natural topsoil from off-site is rapidly becoming a practice of the past. Sources are becoming increasingly scarce. In addition, there is

Table 2. Tree and soil specifications for Harvard Yard restoration

TREE SPECIFICATIONS
• Trees grown in the nursery under long-term contract.
• Wide-spacing with pruning up to 20 feet as height is reached. No less than two-thirds of crown removed during this period.
• Root-pruning on a regular basis. Perhaps use copper-coated root wrap.
SOIL SPECIFICATIONS
• Each tree location excavated for soil inspection.
• Tree-pit design individually specified.
• Underdrainage, etc., installed if necessary.

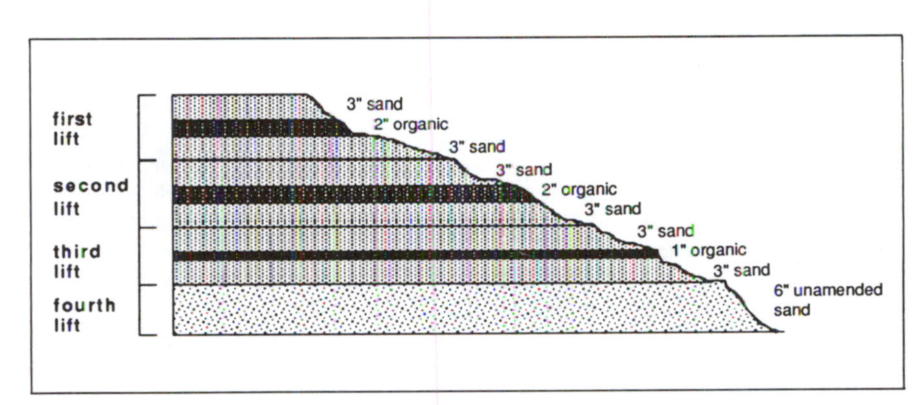

Figure 3 — A constructed profile of till, specified sand and composted sludge.

increasing public sensitivity to the non-renewable use of natural resources and the accompanying environmental impact of creation of stripped soil areas, gravel and sand pits and other landscape disturbances. As Spirn (5) points out so well, we are going to be more involved with the restoration of urban sites with existing unfavorable conditions and we'll have to make do with the materials at hand. This challenge calls for some creative approaches to the solution of designed soil. The design includes both the components used in the soil material and how the profile is constructed — admittedly, not an easy task.

Some years ago, Jim Patterson of the National Park Service was called upon to restore a site on Federal Reserve land that had been a construction staging area for the District of Columbia Metro. Instead of importing a huge amount of topsoil and laying sod, the Park Service tilled the soil, spread a 3-inch layer of composted sludge, tilled it and spread a second 3-inch layer of compost and tilled it. The area was then seeded and rolled. The results were excellent. Hence, Jim and his crews actually created their own topsoil.

Others have now begun to take this idea further, by mixing composted sludge with sand to create a "planting loam" or artificial topsoil. The subsoil is comprised of a mix of specified sand mixed with glacial till and composted sludge. In one particular project, 187,000 cubic yards of designed topsoil and 395,000 cubic yards of designed subsoil are

Table 3. Urban soil component substitutes

Component
MATRIX COMPONENT
• Sand
• Glass
• Zeolites
ORGANIC COMPONENT
• Composted sludge
• Sludge pellets
• Brewer's waste
• Others
PORE-SPACE COMPONENT
• Pumice
• Isolite
• Sintered flyash
• Wicking and vertidrains

required. Even a sand-compost mix requires a huge amount of sand; leaving one or several very large holes in the landscape. Are there other materials that could be used as artificial soil material? Ground glass as a substitute for sand comes to mind, but how about the other components? In an attempt to focus some attention on potential soil substitutes the following list is provided in Table 3. The sand or glass may substitute for the range in sand particle sizes, and perhaps glass could be processed into silt-sized particles. Substitutes for clay may include zeolites or durable synthetic exchange resins.

The organic component may include the better known composted sludge, sludge pellets, brewer's waste, shredded bark, etc. There may be others. With regards to the sludge products, the greatest obstacle is the selling of the idea of spreading "sewage" over the landscape. Obviously, public education is needed for the immediate future. Most vat-processed sludge will meet the EPA requirements of Type 1, which has few environmental limitations. Type 2 is constrained by storage limitations and spreading on steep slopes. Future improvements in the products may decrease their limitations. Brewer's waste sometimes has a residual odor but can be removed with composting; otherwise, it is very satisfactory as an organic additive. In all cases, availability and transportation costs are barriers to wide-spread use of some of these products.

The pore-space component may appear to be a strange inclusion. The soil in the physical sense is above all a porous medium. We must make an effort to consciously provide for a stable pore space through which roots may penetrate, diffusion of gases occurs and yet water is retained in available form. The larger pores contribute to the drainage of the profile under saturated conditions when an impeding layer is not present. The pumice, sintered flyash and Isolite internally provide fine pores for water retention and the larger pores exist as spaces between the larger particles. Cost and availability coupled with transportation of several of these products have prevented more widespread use. The use of VAPAM and other artificial aggregating agents have not met with favor after the rush of research carried out in the 1950's and '60's. The mixes employed by golf course designers and those used in greenhouse flats and pots are too costly for general landscape use. Vertidrains provide artificial saturated drainage pathways, while wicking provides a continuous pathway for the capillary rise of water as a passive irrigation technique. Obviously, the construction of the two products are quite different.

Research Considerations

Future development of our knowledge of urban soils will derive from research that specifically addresses urban soil problems. Up to the present time much of the knowledge has evolved from the application and/or the extrapolation of knowledge from agricultural and silvicultural studies. Unfortunately, this work must be accomplished with decreasing availability of funds and perhaps more restrictions on how the work is accomplished. Therefore, we must make our research efforts more efficient by more sharply focusing on those topics that would have the greatest advancement in urban soil technology. Several of these topics are suggested here.

Heat Budget of Urban Trees and Recognition of Induced Stress. A closer scrutiny of the heat budget (energy balance method) of the tree or a group or trees may be an approach to an evaluation of evapotranspirationon a particular site. This could lead to a clearer picture of the type of soil and the volume required to maintain a tree from one precipitation event to another. This is an indirect approach since there is real difficulty in determining the actual water utilization of individual trees under urban conditions. A suggested initial step is to recognize the various heat stress conditions present as illustrated in Fig. 4.

The single, open tree (Fig. 4a) is exposed to solar radiation but also to a flow of air that tends to reduce crown temperatures and reflected heat from the low-albedo grass surface. The air flow removes moisture from the crown increasing potential transpiration. This is a relatively cool situation. The single enclosed tree (Fig. 4b) is exposed to direct solar radiation in addition to reflected radiation from the pavement below the crown and the building facade. Heat is probably transported by turbulent air from the street, which, more than likely has higher temperatures than the enclosure. Crown temperatures are quite high on a sunny day. Even soil temperatures could reach high levels here. Reduction of temperatures by meso-scale windflow is probably minimal. The copse (Fig. 4c) has a closed canopy, shading the surface below the crowns. Energy exchange (the boundary layer) occurs in the upper crowns rather than around the entire crown as in the single tree situation. Crown temperatures are reduced accordingly. If in a natural setting the litter acts as a mulch, cooling the soil. The trunk space has lower air flow than above the canopy, but the temperatures are moderated. The stress on each tree is quite different for each situation. therefore, the underground landscape needs to be different for each. The details for each situation require elucidation. For the moment we employ our best professional judgement and hope for the best!

Moisture stress may be detected with color infra-red photography. In these images, healthy tree foliage appears as a bright magenta whereas stressed trees are gray. Those moderately stressed are partially gray. Unfortunately, color infra-red film is expensive, unstable (needs to be frozen in storage), very light sensitive and very expensive to develop. In addition, the author is aware of only one laboratory in the entire United States that develops the film (at $20 a roll!). Perhaps a better photographic method of stress detection will be developed in the future.

Dynamic Water Relations of Designed Soil. We still do not have firm ideas about the retention and flow of water in the urban soil profile other than those that are similar to natural profiles. Those that are disturbed or constructed by design have hydraulic properties that may not permit extrapolation of water relations from other known soils. Thus, constructed soils must be studied. For the present, valid application of general soil physics knowledge must be applied to the urban situation with the understanding that not all conditions can be described or resolved. Some questions we may propose

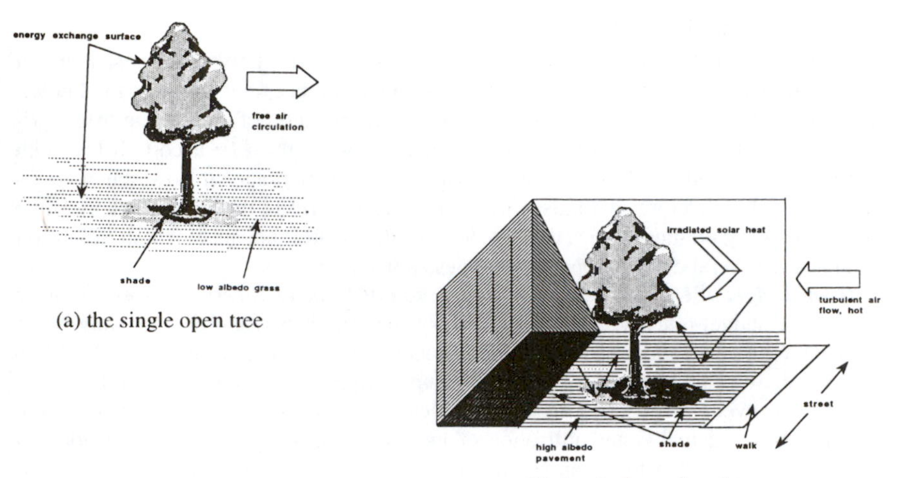

(a) the single open tree

(b) the single enclosed tree

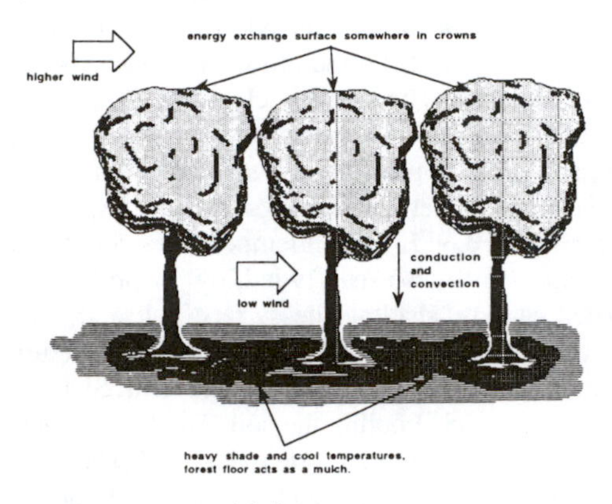

(c) the copse

Figure 4 — Three heat stress situations of the urban tree.

include: just how fast do the roots in a confined space dry out the soil to near the wilt point; or what is the hydraulic conductivity of the several layers (horizons) in the constructed profile; or how does the wetting front, if it exists, move down through the profile when irrigation or precipitation occurs? There are many other unasked questions.

More Passive and Simpler Irrigation Systems. If irrigation is absolutely necessary in the design, then it behooves us as concerned scientists to reduce the dependence of urban vegetation on irrigation with a resource that can become scarce or restricted in any portion of the United States. Since many irrigation systems are subject to damage or mechanical failure, less complex and physically remote systems become more appropriate designs. If we do a better job of matching species cultivars to site conditions and design the underground landscape properly, perhaps the need for irrigation systems can be greatly reduced.

Conclusions

The urban soil situation continues to challenge our scientific expertise and inquiry. The challenge is a dichotomous one. On the one hand we have to apply our present imperfect technology to the real everyday problems faced by the landscape architects, arborists and contractors. Our best may not be good enough. On the other hand, there are real gaps in our knowledge that must be filled with some very practical research coupled with a healthy dose of basic theoretical research on tree (and root) physiology and genetics and their interactions with soil characteristics. Unfortunately, the scientific crossover is difficult to bridge; those interested in urban soils may not be soil scientists, tree physiologists or geneticists or arborists, and the latter group along with landscape architects usually lack training in any of these. The situation is improving as exceptions to the preceding statement increase. The research produced should hold real promise for improved techniques and designs.

Suggested research is costly in time (effort) and manpower. Root research is tedious, usually of destructive nature and not widely popular. Actual on-site urban study is difficult due to exposure of instrumentation or treatments to vandalism or uncontrolled conditions, and resistance by public officials and property owners because of liability questions.

Literature Cited

1. Craul, P.J. 1992. Urban Soil in Landscape Design. John Wiley and Sons, New York. 396 pp.
2. Brewer, R. 1964. The Fabric and Mineral Analysis of Soils. John Wiley and Sons, New York. 470 pp.
3. Harris, R.W. 1991. Arboriculture: Integrated Management of Landscape Trees, Shrubs, and Vines. 2nd ed. Prentice-Hall, Englewood Cliffs, NJ. 674 pp.
4. Hillel, D. 1980. Fundamentals of Soil Physics. John Wiley and Sons, New York. 413 pp.
5. Spirn, A.W. 1992. Designing with the land. J. Soil and Water Cons. 47(1):35-38.

Soil Compaction and Site Construction: Assessment and Case Studies

John M. Lichter and Patricia A. Lindsey

Soils are routinely compacted during the site construction process. This compaction alters the physical properties of these soils, negatively impacting tree preservation and establishment efforts. In order to identify the causes of soil compaction and to assess the severity of compaction that occurs during the various stages of site construction, a study of several residential construction sites in Northern California was initiated. The construction process generally fell into two categories: mass grading, where the entire site is uniformly graded and compacted and selective grading, where either no grading occurs or grading occurs over portions of the site. Bulk densities in landscape areas on a mass graded site in Davis, CA averaged 1.75 g/cm^3. For fenced, undisturbed areas at Briggs Ranch, a selectively graded site in Folsom, CA, bulk densities averaged 1.27 g/cm^3, while outside of fencing, bulk densities averaging 1.61 g/cm^3 were found. On another selectively graded site in Bear Valley, CA, bulk densities in undisturbed soils averaged 0.95 g/cm^3, compared to 1.14 g/cm^3 in disturbed areas. On the first two sites, bulk densities were higher than established critical values, indicating growth limiting conditions on these sites. At Bear Valley, bulk densities in disturbed areas were not above these critical values, which may be a reflection of the soil type present on that site. On mass graded sites, managing compaction requires ameliorative treatments designed to break up compacted soil layers. Soil compaction on selectively graded sites may be significantly reduced through various compaction prevention measures.

Soils are routinely compacted by pedestrian and vehicular traffic and by equipment used during the construction of buildings and their associated landscapes. Forces generated by these events increase soil density, altering soil structure markedly. During compaction, aggregates are destroyed and soil particles are resorted due to both compressive and vibrational forces. This leads to a reduction in the large (macro) pores in the soil, which reduces soil air and water movement (10). In addition, the strength of the soil also increases (1).

Establishing landscapes or preserving trees on these soils following site construction is problematic because these and other alterations render compacted soils less suitable

---

John M Lichter was a Research Horticulturist with the US Forest Service, Center for Urban Forest Research, and is currently Research Associate with the Department of Environmental Horticulture, University of California, Davis, CA 95616-8587. Patricia A. Lindsey is a Landscape Specialist, Department of Environmental Horticulture, University of California, Davis, California 95616-8587.

for plant growth. The most important consequences of these soil physical changes relate to oxygen availability, drainage and soil strength. Reduced oxygen diffusion and drainage in these soils leads to decreased root respiration and slowed root growth. The increased soil strength caused by compaction further limits root growth and function in these soils; limiting drought tolerance, nutrient uptake and plant growth under these conditions (5).

Materials and Methods

In order to further identify the causes and severity of soil compaction during the entire site construction process and to identify where the landscape professional might intervene to mitigate this problem, case studies of several residential construction sites in Northern California were initiated. In addition to observations and interviews with development and construction personnel and landscape professionals, an analysis of soil compaction on these sites was conducted.

In these studies, soil compaction was measured directly as soil bulk density. Bulk density is defined as the dry weight of soil divided by its volume, and is often reported in g/cm^3 or Mg/m^3. In our site analysis, bulk density was measured with a core sampler, a device that is hammered into the soil to the depth of the top of the cylinder, extracting a minimally disturbed sample of soil. Upon drying the sample in a 105°C (220°F) oven for 24 hours, the soil is weighed. This weight divided by the volume of the cylinder (182.8 cm^3) provides the bulk density of the soil in question.

Bulk density results are presented as the average of 5 to 8 samples for each study location. Due to the potential amount of variability in soil physical properties on these disturbed sites, these values offer only a preliminary characterization of soil compaction in these developments.

Results

The first site examined, Northstar Development, in Davis, CA, is a typical high-density residential development in the Central Valley of California. Due to flood mitigation efforts, roadways must act as channels for 100 year floods, therefore, one of the first actions taken was to rough grade the entire site so that lots were 1-2 feet above grade. Residential lots were filled to depths ranging from 0 to 4 feet with material from roadway cuts. In conjunction with the grading process, the entire development site was compacted with specialized equipment to densities required by city ordinance (90% of maximum, City of Davis, CA). This process involved the use of a sheepsfoot roller, which generates extremely high pressures due to its weight and its small area of contact with the soil (the “feet”). In addition, some sheepsfoot rollers vibrate the soil with hydraulic systems aiding further soil compaction (4). The equipment is run when soil conditions are moist and most favorable for compaction. Optimal soil moisture levels are established through laboratory testing. Soils are compacted in 6 to 18 inch layers or “lifts” (4) until the final grade is reached. The above operations prepare individual lots for sale as “finished lots” to contractors, who will then lay foundations (exclusively cement slab), install utility hookups, build the residence, install driveways and other paved areas. Finally, the landscape contractor installs the landscape and the house is placed on the market.

Our bulk density analysis of the back yards of several Northstar finished lots revealed extremely high bulk densities at this stage of development. Bulk densities of the silty clay loam surface soils ranged from 1.63 to 1.82 g/cm^3, with an average of 1.75 g/cm^3. An analysis of soil bulk density in these areas following landscape installation revealed similar bulk densities (1.67 to 1.77 g/cm^3, average, 1.71 g/cm^3). These results suggest that on mass graded sites, the majority of soil compaction occurs during the rough grading/compaction process. The bulk densities determined on this site were well above 1.40 g/cm^3, which has been determined as limiting to plant growth for silty clay loam (3). Soil compaction is likely to be severe and uniform on any mass graded site due to legislation requiring bulk densities between 90-95% of maximum for a particular soil. With the majority of soils, 90% compaction translates to bulk densities that exceed critical values (see Table 1).

Another residential development site, Briggs Ranch, located in the foothills of the Sierra Nevada, in Folsom, CA was designated as a study site. The construction sequence on this site differed from the previous site in several important ways and was typical of many "custom home" developments in this area which are located on slopes with existing woodland or forest. On the majority of the residential lots, foundations utilized discontinuous footings, such as stem walls and piers, which do not require grading or compaction. However, in a few cases where continuous concrete slab foundations were used, grading and compaction was limited to the pad location and occurred prior to foundation installation. Therefore this construction type was termed "selectively graded." After building foundations are installed and utility hookups complete, the construction process continues in a similar fashion to the "mass graded" sites.

While specialized compaction equipment is rarely utilized on these selectively graded sites, soils are compacted by construction equipment and vehicles, as well as from stored materials on these sites. A separate study found that eight passes with a front end loader, which are commonly used on selectively graded sites, raised the bulk density of a silty loam soil from 1.42 to 1.59 g/cm^3 (6). The existence of fenced areas under the canopy of trees to be preserved provided an opportunity to sample the loam soils that were undisturbed for comparison with soils impacted by the construction processes described above. Bulk densities in these fenced areas ranged from 1.05 to 1.42 g/cm^3 (average 1.27 g/cm^3), while densities encountered in non-graded areas outside of fencing averaged 1.61 g/cm^3 (range: 1.56 to 1.70 g/cm^3). Since 1.60 g/cm^3 is a critically limiting bulk density for a loam soil (3), unprotected soils on this site are often limiting to plant growth.

Table 1. Dry bulk density of several soils (g/cm^3) at 85-100% relative compaction as determined by the Proctor Method*

Soil Type	85%	90%	95%	100%	Critical Bulk Density
Well graded loamy sand	1.85	1.96	2.07	2.18	1.75
Well graded sandy loam	1.74	1.85	1.95	2.05	1.70
Med. graded sandy loam	1.64	1.74	1.83	1.93	1.70
Lean sandy silty clay	1.56	1.66	1.75	1.84	1.50
Loessial silt	1.45	1.53	1.62	1.70	1.40
Lean silty clay	1.49	1.58	1.66	1.75	1.40
Heavy clay	1.40	1.49	1.57	1.65	1.40
Very poorly graded sand	1.37	1.45	1.53	1.61	1.75

*Maximum bulk densities from (4) and critical bulk densities from (3).

The final site studied was a resort community in Bear Valley, California, located at 7,000 feet elevation in the Sierra Nevada mountain range. Construction practices were similar to those at Briggs Ranch. Grading was non-existent in building lots and similar machinery was used. Foundations were discontinuous, using several footings per home which varied in thickness from 3 to 5 feet and ran the width or length of the house. A sampling of densities within 10 feet of representative residences revealed bulk densities which averaged 1.14 g/cm^3 (range: 1.00 to 1.28 g/cm^3). Soil bulk densities in undisturbed soils adjacent to these sites (average: 0.95 g/cm^3, range: 0.73 to 1.14 g/cm^3). While bulk densities were increased by the construction process, none of the densities recorded was near the value considered limiting to plant growth. This may be attributed to the soil type present on the test sites (a loamy sand). Soils such as these, with narrow particle size distribution, are less compactible than those with a wide distribution of particle sizes (4). Therefore, a knowledge of soil particle size distribution is helpful in predicting changes in bulk density resulting from compaction of different soil types.

Summary

Successful compaction management requires a thorough understanding of the construction process. The above descriptions of residential site construction in California indicate that construction processes differ among developments, but fall into two general categories: mass graded and selectively graded. On mass graded sites, preventing soil compaction would involve eliminating the grading and compaction process; an integral part of this type of construction. Therefore, compaction prevention on these sites is largely impractical. However, by fencing off sensitive soil areas (which include areas with existing tree roots), it may be possible to achieve this end. Because soil compaction is a limiting factor to plant growth on these sites, remediating these compacted soils should be attempted. This treatment involves soil bulk density analysis to establish the severity and location of compacted soil zones (usually deeper than fill depth) following construction, remediative treatment prior to landscape installation, utilizing deep tillage or subsoiling below the compacted layer (7,8,9), and evaluation over at least two growing seasons. The cost of these procedures is often high and information regarding their effectiveness in urban areas is limited.

In contrast to mass graded sites, soil compaction may be prevented effectively on selectively graded construction sites. By siting infrastructure sensibly, minimizing graded areas and identifying vehicle access and material storage areas during the planning stages of a development, the area of compacted soil may be minimized. Fencing sensitive tree root zones or other areas on a site will eliminate compaction in localized areas. In addition, the use of protective surface treatments such as 6" of mulch or 4" of gravel may reduce soil compaction from vehicle traffic in unfenced areas (6). Finally, restricting traffic and soil work to periods when soils are dry will also help to prevent soil compaction (2). The use of the above techniques should help to reduce soil compaction on a site and eliminate or reduce the need for remediative techniques on these sites.

In conclusion, soil compaction is of great concern for the growth of recently planted and existing vegetation in newly developed landscapes. Successful management of soil compaction is only possible with a thorough knowledge of the site construction process.

Literature Cited

1. Alberty, C.A., Pellett, H.M. and Taylor, D.H. 1984. Characterization of soil compaction at construction sites and woody plant response. J. Environ. Hort. 2(2):48-53.
2. Craul, P. 1992. Urban Soil in Landscape Design. John Wiley and Sons, Inc. New York, USA, 396 p.
3. Daddow, R.L. and Warrington, G.E. 1983. Growth limiting soil bulk densities as influenced by soil texture. US Forest Service, WSDG Report, WSDG-TN-00005, 17 p.
4. Holtz, R.D. and Kovacs, W.D. 1981. An Introduction to Geotechnical Engineering. Prentice Hall, Englewood Cliffs, New Jersey, 733 p.
5. Kozlowski, T.T. 1985. Tree growth in response to environmental stresses. J. Arboric. 11(4):97-111.
6. Lichter, J.M. and Lindsey, P.A. 1993. The use of surface treatments for the prevention of soil compaction during site construction. J. Arboric. In review.
7. Morris, L. 1988. Influence of site preparation on soil conditions affecting stand establishment and tree growth. Southern J. Appl. Forestry 12(3):170-178.
8. Oster, J., Singer, M., Fulton, A., Richardson, W. and Prichard, T. 1992. Water Penetration Problems in California Soils. Kearny Foundation of Soil Science. 165 pp.
9. Rolf, K. 1991. Soil improvement and increased growth response from subsoil cultivation. J. Arboric. 17(7):200-204.
10. Ruark, G., Mader, D. and Tattar, T. 1982. The influence of soil compaction and aeration on the root growth and vigor of trees-A Literature Review. Part 1. Arboric. J. 6:251-265.

Soil Compaction and Loosening Effects on Soil Physics and Tree Growth

Kaj Rolf

The effects of subsoil compaction, topsoil compaction, mechanical subsoiling with an excavator and pneumatic subsoiling using a Terralift soil aerator on a number of soil characteristics were measured at two sites, one with sandy loam and another with loam over clay loam. Soil compaction was undertaken under controlled conditions prior to subsoiling. Compaction had a negative influence on soil physical properties and plant growth. Subsoiling with an excavator had a good loosening effect on the soils and increased growth on the sandy loam. The soil aeration treatment did loosen the sandy loam but not the clay loam. No growth increase was observed.

Soil compaction has a strong influence on the growth and character of vegetation. During building operations, soil is compressed, either intentionally or unintentionally, and may result in long term changes in the soil structure due primarily to the use of heavy vehicles or machines under unsuitable conditions.

In order to create good growing conditions some kind of soil loosening must be undertaken. In agriculture, subsoilers of different types have been used with varying degrees of success, but on construction sites this type of equipment is often ineffectual since tillage aimed at improving the soil structure is carried out when the soil is not dry enough to obtain optimal results.

Macropores are essential for the movement of gases and water in the soil, but when compacted these large pores are destroyed (24). Without adequate pore space between soil particles there is a risk of anaerobic conditions in the soil, the negative effects of which can have an adverse effect on the growth of plants (9,10,12,13,21,23). Well aerated soil structure and texture are as essential as water and nutrients for most plants. An exact value for the critical oxygen content in the soil is very difficult to determine since this not only depends on soil structure and pore sizes but also on the demands of the roots. Factors determining these demands include genetic background and age of root tissue, radius of the root and diffusion coefficient within the root (14). These factors will cause variations in the critical oxygen demands depending on variations among species.

Excavators for subsoiling have successfully been used in fruit orchards in the United States (8) and trials have also been carried out on cultivated soils in Sweden. Håkansson (11) has reported important effects on drainage conditions even though this is a very complex problem. The effect on drainage conditions in the soil depends on the thor-

---

Kaj Rolf is with The Swedish University of Agricultural Sciences, Department of Agricultural Engineering, Section of Horticultural Engineering, Box 66, S-23053, Alnarp, Sweden.

oughness of the loosening of the soil, on precipitation, on evaporation and drainage etc. Positive results can be expected in compacted/impermeable soils.

Compressed air has been used to alleviate poor soil conditions for at least sixty years (22) and Yelenoski (25) reported experiments using portable air compressors from which it was concluded that the release of compressed air into the soil improved soil aeration under certain conditions.

The Terralift equipment was introduced in the late seventies. The unit consists of a petrol engine, which powers a compressor. The whole unit is run and operated by compressed air and pneumatic motors. A probe is air-hammered into the soil and compressed air is released through the probe with a pressure of up to 20 bars (290 psi). The air is assumed to fracture the soil and at the same time it is possible to inject dry fertilizer, lime or dry material that stabilizes the fractures and raises the soil's nutrient level.

Experiments were undertaken to determine the extent of soil compaction on construction sites and to test the ability of an excavator and Terralift equipment to alleviate the conditions described above.

Materials and Methods

Two sites—a sandy loam at Alnarp and a loam over clay loam at Landskrona were chosen for a controlled experiment in May, 1987. The topsoil was removed to a depth of 0.3 m and the subsoil compacted with a wheeled excavator which drove over the site ten times when the soil was dry. The excavator had two axles, an axle load of 8 Mg, and tire inflation pressure of 500 kPa. After the compaction the topsoil was returned (compacted control plot C and H). The treatments are summarized in Table 1.

One plot at each site was subsoiled (plots A and F), directly after the above mentioned treatment, using an excavator, the operation of which is illustrated in Figure 1. The soil was not turned over - it was allowed to fall back into place, but compacted soil layers were broken up thereby creating cracks in the soil.

The breaking up process described by Håkansson (11) was carried out with the topsoil removed, while in these experiments top soil was replaced on the subsoil before treatment. Whichever method is used, a certain amount of topsoil will fill the cracks created when the soil is lifted. These tilled cracks are excellent paths for root penetration since the soil is loose and the organic content provides nutrients.

One plot at each site (plots B and G) was treated, after the above mentioned compaction treatment, with the Terralift soil aerator (Fig. 2). The probe was driven into the soil on 1 meter surface grid spacing and compressed air was discharged at two depths, 0.45 and 0.75 meter.

The surface compacted plots (D and I) were compacted with the same machine and intensity as was the case with the subsoil compaction, but here the load was applied on the topsoil.

The untreated control plots (E and J) were normal arable land with a plow pan.

Due to practical and economic reasons only one plot with each treatment was laid out at each site. Each plot was 5 x 13 m, where half the area was planted with trees and the other half (with annual agricultural crops) was used for soil physical examinations in order to avoid disturbance of the tree root systems.

The treatments were applied to five species, *Acer platanoides*, *Corylus avellana*, *Fraxinus excelsior*, *Sorbus intermedia* and *Quercus robur*. Five plants of each species were planted on 1 meter grid spacing in each plot.

Table 1. Plot identifications of the different treatments at the two sites, Alnarp and Landskrona

Plot	Place	Plot	Place	Treatment
A	Alnarp	F	Landskrona	Subsoiled with excavator
B	Alnarp	G	Landskrona	Terralift treated
C	Alnarp	H	Landskrona	Compacted control
D	Alnarp	I	Landskrona	Surface compacted
E	Alnarp	J	Landskrona	Untreated control

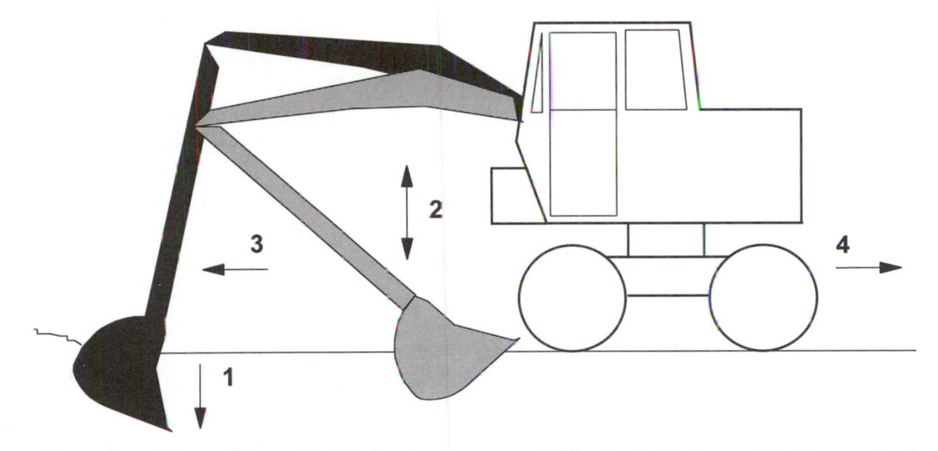

Figure 1 — The working principle for the excavator. 1) The bucket is lowered to the required depth. 2) The bucket is lifted up and shaken. 3) The soil is dropped back into the hole. 4) The excavator reverses.

Figure 2 — The Terralift equipment at the university experimental fields.

Cores of soil (72 mm in diameter and 100 mm long with 6 - 10 replicates for every depth) were collected to 0.6 m depth 6 and 18 months after treatments for determination of bulk density, pore volume, pore size distribution, saturated hydraulic conductivity and air permeability. Saturated hydraulic conductivity, Ks, was measured after saturating the soil cores and applying a constant-head. The apparatus used is described by Andersson (1). The soil water-release characteristic was determined using standard methods (3,17) at water matric tensions of -1.5 kPa using a tension table and at -10 kPa and -60 kPa using pressure plate apparatus. Total porosity was calculated from the particle and dry bulk densities. Air filled porosity was calculated as the difference between total porosity and the volumetric water content at the different tensions. The air permeability coefficient, K_a, was measured on the soil cores, according to Andersson (2) in which the volume of air drawn through the soil sample at a known potential is measured for a 120 second period.

Bulk densities were calculated after oven drying to constant weight, at 105°C. Field saturated hydraulic conductivity, K_{fs}, was measured using double ring infiltrometers of 0.40 m (inner ring) and 0.63 m (outer ring) diameter and 0.25 m height (4). Rings were installed in the soil to a depth of 0.1 m. Steady state flow was assumed to occur after 2-2.5 hours of infiltration.

Penetration resistance (cone pressure) was measured with an Electronic cone penetrometer, constructed at the department (15). Cone pressure is the instantaneous penetration force divided by the cone base area. Data were collected for every 10 mm level with 30 replicates for each plot. An average was calculated for each level.

Every year in October total shoot extension was recorded. After 4 years, total plant heights were measured and plants at the Alnarp site were dug up to record root morphology. Plants from plot E were not dug up because of frozen soil. The weight of the above ground part of the plants was calculated after oven drying, at 105°C until constant weight.

T-tests were used to determine statistical significance of all data at the 95 percent confidence level of significance. Because of the experimental design this test only tells if there were differences between plots but not the reason for the differences. This paper discusses the possible reasons.

Results and Discussions

In this paper the results of soil physical properties of the surface compacted and untreated control plots are given. Soil physical changes due to the subsoiling and Terralift treatments are reported in Rolf (19) and Rolf (20) and a summary of the results are given below.

Subsoiling with the excavator resulted in a significantly reduced soil bulk density. Pore volume increased up to 42 percent and airfilled porosity at field capacity increased as well. Penetration resistance was lowered for both sites (A and F).

The effects of pneumatic subsoiling using the Terralift soil aeration equipment, for the Alnarp sandy loam, were a decreased bulk density, an increase in porosity, mainly as an increase in macroporosity and an increase in saturated hydraulic conductivity and air permeability. Penetration resistance was lowered at this site. After treatment of the loam soil at the Landskrona site, bulk density increased and porosity decreased. There were no changes in saturated hydraulic conductivity and a small increase in air permeability. The second year after treatment an increase in macroporosity was observed. Penetration resistance was higher the first year, as could be expected from the bulk densities, but not different from the compacted control 2 years later.

Dry Bulk Density

Mean bulk densities were significantly higher in the surface compacted treatment at Alnarp, six months and eighteen months after treatment, compared to the untreated control.

At the Landskrona site the results were the same as Alnarp down to 0.3 m depth. Below there was no significant difference in 1987, however in 1988 a difference appeared for the 0.3-0.4 m layer as shown in Table 2. This difference is suspect since there should have been a change in the control from 1.63 to 1.52 Mg m-3 from 1987 to 1988 and this cannot be true. This response can only be attributed to variation within the plot.

Porosity and Pore Size Distribution

At the Alnarp site, the surface compacted plot had significantly lower porosity values than the untreated control and a different pore size distribution. The values were not changed from 1987 to the next year, indicating that the natural processes in the soil did not loosen the soil to any measurable extent from this year to the next. Since porosity values (Table 3) were calculated from bulk densities there were significant porosity changes at the Landskrona site in the same layers as was the case with bulk density.

The Alnarp site had lower macroporosity (Table 4) in the surface compacted plot compared with the untreated control except in the 0.2-0.3 m layer. This exception can be due to the already existing plow pan where the soil was already compacted to a certain extent before the experiment was set up. This plow pan can easily be identified in the penetration curve in Figure 3.

At Alnarp the percentage of medium sized (0.03-0.005 mm) pores was significantly lower in the surface compacted plot compared to the untreated control (Table 5) in 1987, except for the 0.4 - 0.5 m layer. There was no difference in percentage of fine pores. The change in porosity was a change in percentage of macro and medium sized pores.

In 1988 the results were the same except for the 0.2 - 0.3 m layer where there was a

Table 2. Mean bulk densities, in Mg/m^3, 6 months and 18 months after treatment. n = 6-10

Site	Depth m	6 months		18 months	
		Control	Compacted	Control	Compacted
Alnarp	0.0-0.1	1.58	1.63*	-	-
	0.1-0.2	1.65	1.80*	1.65	1.80*
	0.2-0.3	1.75	1.79*	1.72	1.78*
	0.3-0.4	1.64	1.75*	1.62	1.74*
	0.4-0.5	1.57	1.62*	1.57	1.63*
Landskrona	0.0-0.1	1.45	1.64*	-	-
	0.1-0.2	1.35	1.67*	1.31	1.69*
	0.2-0.3	1.41	1.65*	1.45	1.71*
	0.3-0.4	1.63	1.66	1.52	1.71*
	0.4-0.5	1.58	1.66	1.67	1.67

* = Statistically significant from control ($P \leq 0.05$)

Table 3. Porosity (percent v/v), 6 months and 18 months after treatment. n = 6-10

Site	Depth m	6 months Control	6 months Compacted	18 months Control	18 months Compacted
Alnarp	0.0-0.1	39.1	37.7	-	-
	0.1-0.2	36.3	31.5*	36.4	31.4*
	0.2-0.3	33.5	31.7*	34.7	32.2*
	0.3-0.4	38.6	34.4*	39.2	34.5*
	0.4-0.5	41.6	38.8*	41.7	38.4*
Landskrona	0.0-0.1	44.5	37.0*	-	-
	0.1-0.2	48.3	36.2*	49.8	35.5*
	0.2-0.3	45.7	36.8*	44.3	34.4*
	0.3-0.4	38.0	37.0	42.2	35.3*
	0.4-0.5	41.3	38.2	37.9	37.9

* = Statistically significant from control ($P \leq 0.05$)

Table 4. Macroporosity (pores > 0.03 mm) (percent v/v), 6 months and 18 months after treatment. n = 6-10

Site	Depth m	6 months Control	6 months Compacted	18 months Control	18 months Compacted
Alnarp	0.0-0.1	11.4	10.7	-	-
	0.1-0.2	7.3	4.1*	7.2	4.2*
	0.2-0.3	4.6	4.2	5.5	5.2
	0.3-0.4	8.6	5.6*	8.9	4.3*
	0.4-0.5	13.0	10.5*	12.2	7.8*
Landskrona	0.0-0.1	12.7	4.9*	-	-
	0.1-0.2	18.3	2.8*	22.2	2.8*
	0.2-0.3	15.1	4.0*	13.0	1.9*
	0.3-0.4	6.6	6.8	11.8	4.2*
	0.4-0.5	10.5	7.3	11.3	4.6*

* = Statistically significant from control ($P \leq 0.05$)

significant lower percentage of fine pores in the surface compacted plot. The plow pan may be the reason for the different soil reaction but there is no good explanation of this reaction.

At Landskrona macroporosity (Table 4) in the surface compacted plot was significantly lower than in the control plot, down to 0.3 m in 1987 and down to 0.5 m in 1988. Since the Landskrona soil is a swelling and shrinking type of clay the changes in pore size distribution from one year to another may be explained by the fact that there was a dry summer in 1988 and the soil had up to 3 cm wide cracks in the soil surface. The change in the 0.3 - 0.4 m layer from 1987 to 1988 can be due to this fact but can also be a result of the earlier mentioned possibility of variations within the plot.

Saturated Hydraulic Conductivity

Hydraulic conductivity is a measure of a soil's potential for water transport. It is dependent of the macro pore system but also on the continuity of that system. At both

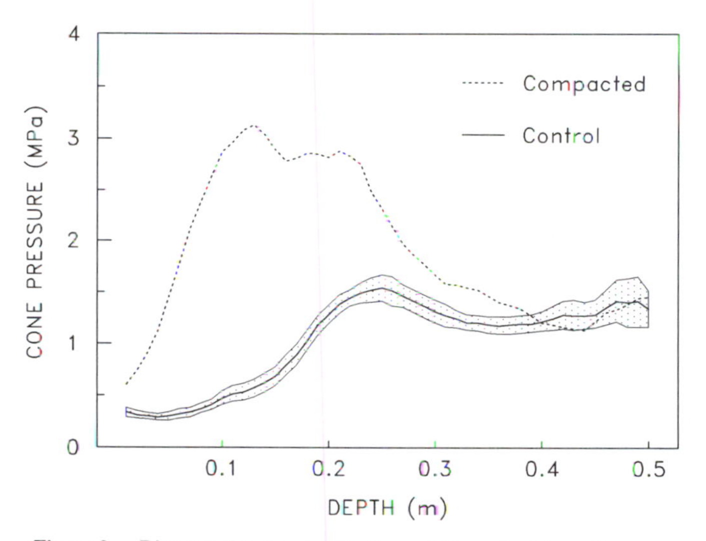

Figure 3 — Diagram showing penetration resistance for the surface compacted ploy and untreated control, at the Alnarp site, 6 months after treatment. Each curve is a mean of 30 individual penetration curves. The control is surrounded by a 95 percent confidence interval based on the pooled variance of all curves.

Table 5. Medium (0.03-0.005 mm) and fine (<0.005 mm) pores (percent v/v), 6 months and 18 months after treatment. n = 6-10

		6 months				18 months			
		medium		fine		medium		fine	
Site	Depth m	Ctrl	Comp	Ctrl	Comp	Ctrl	Comp	Ctrl	Comp
Alnarp	0.0-0.1	4.1	4.0	23.6	23.0	-	-	-	-
	0.1-0.2	3.5	2.2*	25.6	25.2	4.7	3.0*	24.4	24.2
	0.2-0.3	3.3	2.2*	25.6	25.4	3.0	2.7	26.2	24.3*
	0.3-0.4	3.5	3.0*	26.5	25.8	3.8	3.0*	26.6	27.2
	0.4-0.5	3.7	4.6*	25.0	23.7	3.6	4.3*	25.9	26.3
Landskrona	0.0-0.1	3.8	3.1*	28.0	29.1	-	-	-	-
	0.1-0.2	3.6	2.3*	26.7	31.1*	4.0	2.4*	23.7	30.3*
	0.2-0.3	3.6	2.6*	27.0	30.2*	4.0	2.3*	27.3	30.2*
	0.3-0.4	3.0	2.6*	28.4	28.0	3.5	1.9*	26.9	29.2*
	0.4-0.5	3.0	2.6*	27.8	28.2	3.5	3.3	23.2	30.0*

* = Statistically significant ($P \leq 0.05$)

sites the untreated control plots had higher saturated hydraulic conductivity values than the surface compacted plots (Table 6) where there were significant differences in macroporosity. Due to the great sample variation, median values are presented instead of means (5). Infiltrometers measure the infiltration behavior of soils and in particular the soil's infiltration capacity. The infiltration capacity is identified theoretically with the hydraulic conductivity of the saturated soil, so that measurements on infiltration can provide estimates of field saturated hydraulic conductivity values (26).

Field saturated hydraulic conductivity (Table 7) was higher in the control plots than the surface compacted plots at both sites. The range values are separated and verifies the results from the laboratory measurements that there is a risk of slow drainage in the compacted plots.

Soil Aeration

The values in Table 4 are equivalent to airfilled porosity at -1.0 meter water column or at field capacity. Richards and Cockroft (16) found, in a study on apple trees, that root growth was inhibited when airfilled porosity was less than 15 percent. Edling (6) recommended that the volumetric content of airfilled pores should be more than ten percent in order that an adequate level for root penetration is maintained (valid for a topsoil with low clay content). In a clay soil a lower value can be sufficient, in a sand probably a higher, maybe 15 percent, may be needed (I. Håkansson, personal communication).

Both the surface compacted and the control plot at Alnarp had low airfilled porosity down to 0.4 m depth. The higher values in the top layer in 1987 are due to mechanical weed control on the soil surface. There was a significant difference between the treatments except in the 0.2 - 0.3 m layer where a plow pan had been identified.

At the Landskrona site there also was a significant difference between the treatments but here the control plot had values higher than the limit of 10%. In the surface compacted plot there was a risk for oxygen deficiency throughout the whole profile.

The size of the air permeability coefficient depends on airfilled porosity and pore radius and indirectly on soil texture, soil structure, bulk density and soil water status. The air permeability coefficients (Table 8), K_a, (cm/min) followed trends similar to that of airfilled porosity, except in the 0.2 - 0.3 m layer at the Alnarp site where there was a lower median value in the control compared to the surface compacted plot. Air permeability also followed the trends of hydraulic conductivity. At the Landskrona site there was a much higher air permeability in the control plot compared to the surface compacted plot.

Edling (6) notes a guide value of 400 cm/min as the lowest value for air permeability for normal arable land. Only the 0.1-0.2 m layer in the control plot at Landskrona reached this guide line.

Penetration Resistance

A cone penetrometer is a simple tool for assessing the mechanical condition of a soil. In theory, the lower the cone pressure, the looser the soil and the more easily it can be penetrated by the roots. In Figure 3 it can be seen that the curve for the control plot at Alnarp had an increase in cone pressure from 0.1 m to 0.25 m. This peak was due to the compaction in the plow pan. The cone pressure for the surface compacted plot was significantly higher with a maximum at 0.12 m of over 3 MPa, then decreasing down to 0.4 m. This plot was influenced by the compaction down to this depth of about 0.4 m.

At the Landskrona site (Figure 4) the plow pan could be recognized from 0.2 m down to 0.4 m depth in the control curve. The curve for the surface compacted plot show a maximum value for cone pressure at 0.11 m depth and is clearly separated from the control curve down to 0.5 m depth. There was an influence of the compacting treatment down to 0.5 m.

Plant Growth

The experience from a previous study (18) was that after the first year after planting

Table 6. Median values (50-percentile) for the saturated hydraulic conductivity, K_s, (cm/h) for the examined depths, 6 months and 18 months after treatment. n = 6-10

		6 months		18 months	
Site	Depth m	Control	Compacted	Control	Compacted
Alnarp	0.0-0.1	0.69	2.60	-	-
	0.1-0.2	0.69	0.05	0.00	0.02
	0.2-0.3	0.05	0.01	0.02	1.07
	0.3-0.4	0.26	0.00	0.34	0.02
	0.4-0.5	10.24	0.15	0.84	0.05
Landskrona	0.0-0.1	10.69	0.00	-	-
	0.1-0.2	26.69	0.02	27.70	0.03
	0.2-0.3	36.09	0.02	21.80	0.05
	0.3-0.4	0.02	1.87	1.80	0.75
	0.4-0.5	16.29	0.44	15.30	0.76

Table 7. Field saturated hydraulic conductivity, K_{fs}, (cm/h). Values are final results after 2-2.5 hours of flow. Values are from measurements 30 months after treatment, (n=3) at the soil surface

	Alnarp		Landskrona	
	Control	Compacted	Control	Compacted
Median	6.1	1.6	2.2	1.1
Range	3.3-6.6	0.5-3.2	2.1-7.8	0.5-1.6

Table 8. Median values (50-percentile) for air permeability coefficient, K_a, (cm/min) for the examined depths at -1.0 m water column, 18 months after treatment. n = 6-10

	Alnarp		Landskrona	
Depth m	Control	Compacted	Control	Compacted
0.0-0.1				
0.1-0.2	20.0	3.1	2041.0	2.0
0.2-0.3	5.7	14.2	289.7	0.9
0.3-0.4	30.2	1.1	59.0	11.3
0.4-0.5	88.9	2.5	152.5	5.4

there was no real difference in growth due to the limited root system that is not really restricted in development because it has not yet reached the compacted horizon.

There was a similar growth reaction at Alnarp and Landskrona. The second year was very dry and for some species the growth was even better (not significantly) in the compacted control. This may be due to more plant available water in the compacted control, because of a larger amount of small capillary pores. In the subsoiled plot, macropores drain away the water. For a region of low rainfall there is a risk for drought stress in a

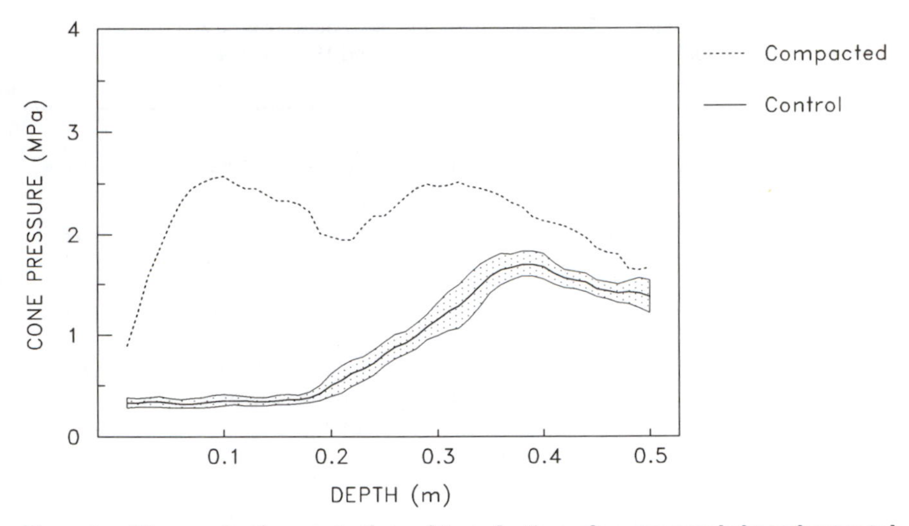

Figure 4 — Diagram showing penetration resistance for the surface compacted plot and untreated control, at the Landskrona site, 6 months after treatment. Each curve is a mean of 30 individual penetration curves. The control is surrounded by a 95 percent confidence internal based on the pooled variance of all curves.

subsoiled area. The dry summer also dried out the clay soil at Landskrona so that the soil shrunk and cracked, which created aeration and rooting paths that may have diminished the differences between treatments.

Top Shoot Growth

At both sites top shoot growth, independent of species (Fig. 5 and 6), was significantly lower in the surface compacted treatment compared to the untreated control and the other treatments. Considering the growth there was no significant difference between the excavator subsoiling and the compacted control, except for Alnarp, 1989. There was no difference between the Terralift treatment and the compacted control, but for 1988 and 1989 there was significantly better growth at the Alnarp site in the excavator treated plot compared to the Terralifted plot.

When the growth figures are separated into each species there were some significant differences between the treatments. In the second and third year of growth at the Alnarp site, there were significant differences for maple and ash between the excavator and the compacted control, and for maple and oak between the surface compacted and untreated control. No differences for the Terralift treatment.

At the Landskrona site the main difference emerged in the third year of growth. For maple, ash, whitebeam and oak there was significantly less growth in the surface compacted plot compared to the untreated control. The Terralift treatment had no significant effect on top shoot growth and the excavator treatment only had an effect on maple the third year.

Total Height

Figures 7 and 8 show the total height of the plants after 4 years. At the Alnarp site there were significant differences in total height for maple, hazel, ash and oak in the

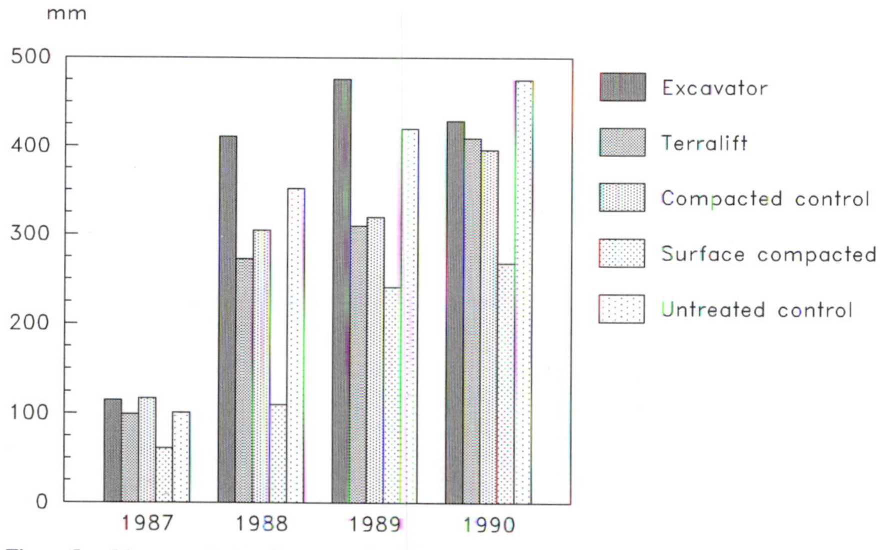

Figure 5 — Mean yearly top shoot growth of plants at Alnarp 1987-1990. n=25.

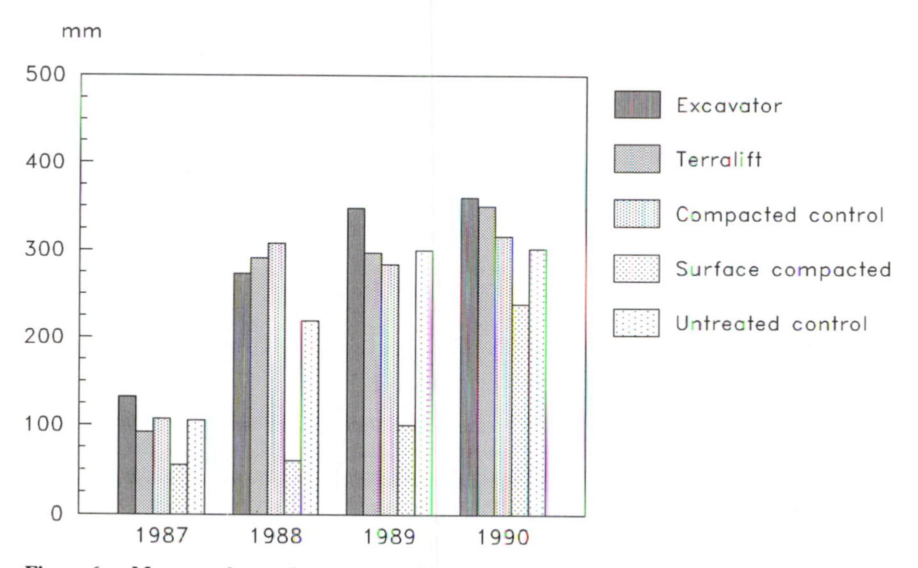

Figure 6 — Mean yearly top shoot growth of plants at Landskrona 1987-1990. n=25.

excavator subsoiled plot, compared to the compacted control. The Terralift treatment did not result in any significant growth differences. All species were significantly higher in the untreated control plot compared to the surface compacted plot.

At Landskrona no treatment had any positive effect on total height. The only result was that hazel, ash and whitebeam in the surface compacted plot had less growth than the untreated control.

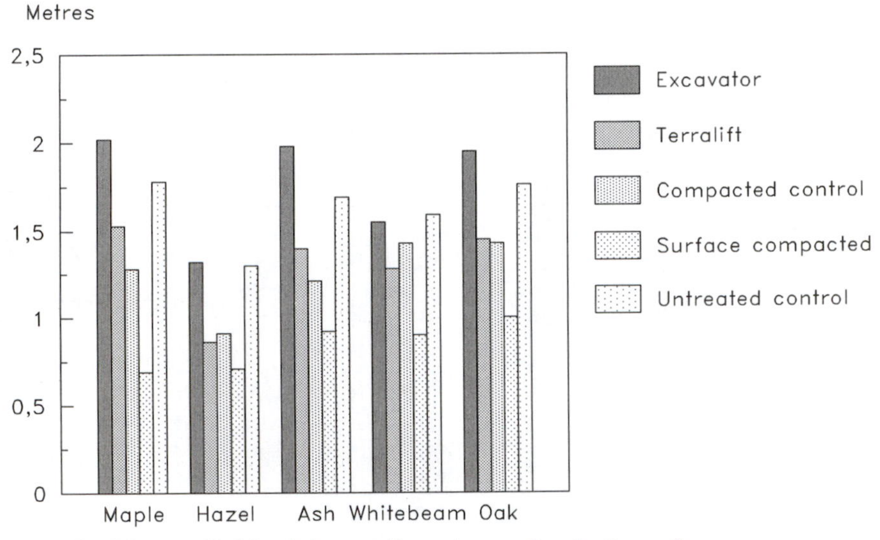

Figure 7 — Mean total height of plants at Alnarp 4 years after planting. n=5.

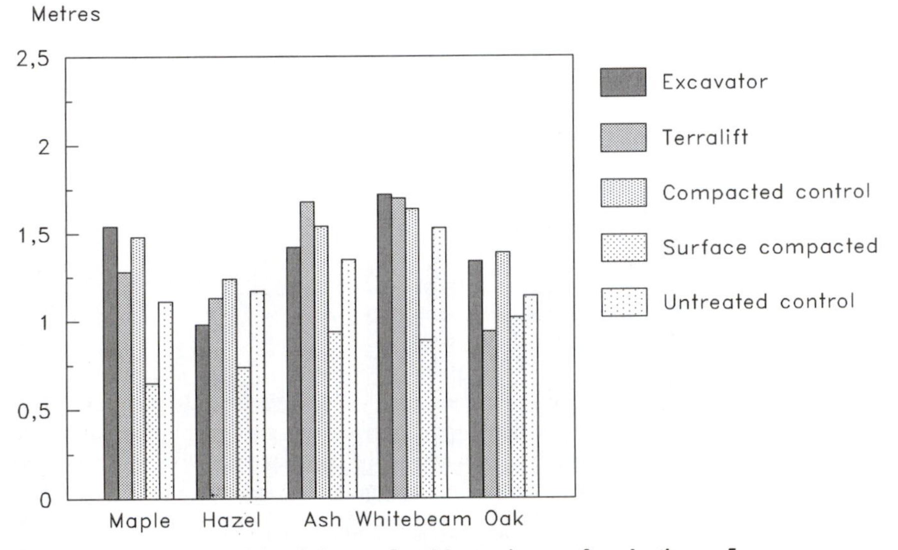

Figure 8 — Mean total height of plants at Landskrona 4 years after planting. n=5.

The growth reaction in the excavator subsoiled plot at Alnarp was expected in accordance with earlier experiments (18), but there was no difference in growth at Landskrona. The soil at Landskrona is a shrinking and swelling type of clay and maybe this type of soil does not need any help with soil loosening since the soil itself naturally aerates, within a short period of time. El-Araby et. al. (7) reported similar reactions from subsoiling experiments on montmorillonitic clays. On the other hand soil physical mea-

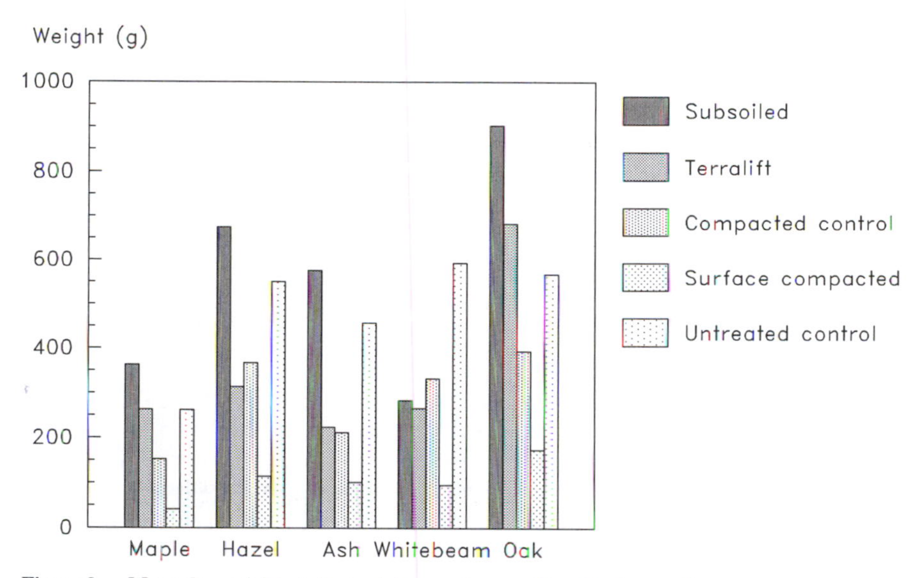

Figure 9 — Mean dry weight of plants at Alnarp 4 years after planting. n=5.

surements at Landskrona indicated problems in the compacted control plot with low values for macroporosity, hydraulic conductivity and air permeability. Perhaps the large cracks as a result of the shrinking can still supply the root systems with sufficient oxygen.

Biomass

Mean dry weight of plant parts above ground had the same significant differences as in the case with total plant height. At Alnarp maple, hazel, ash and oak had more biomass following the excavator treatment than in the compacted control and all species had less biomass in the surface compacted plot compared to the untreated control (Figure 9). No differences were found between plants in the Terralifted plot and the compacted control.

Root Growth

At the Alnarp site root systems were dug up after four years in all plots except the untreated control. No real measurements were done on root growth but a visual inspection of root morphology showed that the plants in the compacted control had more horizontal root systems than the plants in the loosened plots. The plants in the surface compacted plot had small and horizontal root systems. Median plants are shown in Figures 10-14.

Conclusions

An eight ton axle load caused decreases in bulk density, pore volume, and the percentage of macropores. These changes had a negative influence on hydraulic conductivity and air permeability. Plant growth was, to a large degree influenced negatively by these soil physical changes. Growth was slow and the plants did not develop wide and deep root systems.

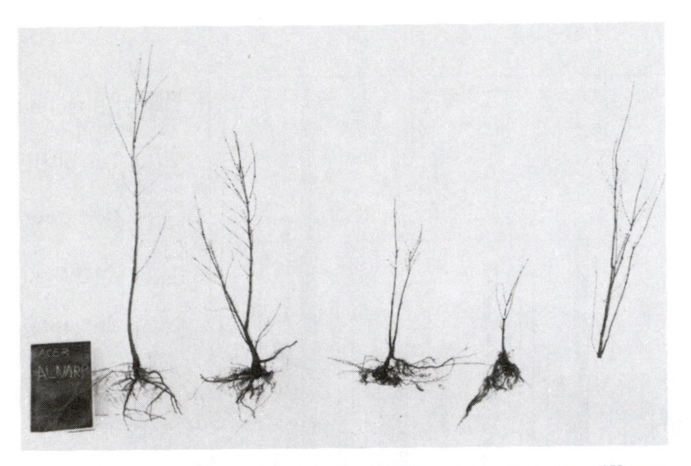

Figure 10 — *Acer platanoides*, Maple. Median plants from the different treatments. From the left plots, A to E.

Figure 11 — *Corylus avellana*, Hazel. Median plants from the different treatments. From the left plots, A to D.

Soil compacted in this way must be treated. The object of using an excavator or compressed air as a tool for subsoiling, is to give plants better growing conditions in the soil and to increase porosity in compacted horizons.

Air permeability coefficient (K_a) is a function of the air content of the soil and shape of the air-filled pores. Subsoiling can create more macropores which can be beneficial but may also disturb the existing pore continuity. Both macroporosity and hydraulic conductivity were higher in the treated plots indicating both volume increase and a continuous pore system.

The water holding capacity was improved where pore volume was increased. The amount of plant available water at field capacity for each horizon was however not increased, since all water from the "extra" pore volume was contained in macropores

Figure 12 — *Fraxinus exelsior*, Ash. Median plants from the different treatments. From the left plots, A to E.

Figure 13 — *Sorbus intermedia*, Swedish whitebeam. Median plants from the different treatments. From the left plots, A to E.

and therefore drains away. However, the better and more uniform porosity through the profile offers the roots the possibility of vertical expansion and access to more water and nutrients.

The penetration resistance can be regarded as a verification of the other measurements made to determine soil physical changes, especially bulk density and pore volume changes.

Plant growth was reduced by the surface compaction at both sites but only the excavator treatment at Alnarp had a positive influence on plant growth. At the Landskrona site, with the clay soil, better growth due to the excavator treatment was expected since an earlier experiment (18) on a soil with similar clay content showed good growth results. No clay mineral analysis was made so there may have been differences in clay

Figure 14 — *Quercus robur*, Common oak. Median plants from the different treatments. From the left plots, A to E.

mineral type. The clay soil at Landskrona had wide cracks in a dry period and appeared to be able to aerate itself naturally in a satisfactory way. Some moraine clays are known to provide good growth responses although a physical examination of these soils suggests a poor growth response. In such instances subsoiling is questionable.

Why then was there no increase in growth with the Terralift treatment at Alnarp? Soil physical parameters were improved nearly as much as with the excavator treatment, but without significant growth response. There were no important differences in soil chemistry between the plots and the soil texture was the same in the Terralifted and compacted control plots. The plot treated with the excavator had a little different texture depending on the soil mixing that occurs when digging, but this can't explain the results. It is suggested that the measuring methods were inadequate to identify the nature of the condition of the soil. The reaction of the roots would appear to be a better measure of the condition of the soil.

The experiments with Terralift were conducted before planting and without any substrate injected in the soil. This is not the normal way to use this kind of equipment. The intention is to use the Terralift in or around existing plantations and trees. This project has mainly concentrated on soil physical effects and results on growth may have been different if the treatment had been done sometime after plants had been established. When Terralift is used in established vegetation there is reason to believe that some kind of root pruning will take place. This is not necessarily negative for the plants. This pruning can stimulate root growth and if soil physics are improved then this can be beneficial for the plants. The subject is now being studied.

There is always a discussion about long term effects of different treatments. This experiment lasted for four years and the fourth year top shoot growth seems to have stabilized on a certain growth level. The best effect of the excavator treatment was the second and third year after planting. There is always a likelihood of recompaction if subsoiled areas are exposed to subsequent loads or vibration. An area which has been subsoiled is much more liable to compaction since the natural structure in the soil has been changed. After a few years, the soil will stabilize or, if the area is subjected to high traffic density, the vibration will create a different pattern of compaction. The long term effect is very much dependent upon what happens after treatment.

The load that the sites were exposed to have had an influence on the macropore system so that the volumetric content of air filled pores is lowered. The same has happened with the pore volume. Important soil physical parameters are negatively affected and when the soil has suffered this compaction damage, as is the case on most construction sites, it must be loosened before planting.

The use of an excavator for subsoiling can be recommended on most soils that need to be treated due to soil compaction. The method is used by contractors both in Sweden and in, for example, USA (Ted Stamen, pers. com.) with good results. The method is especially satisfactory for small and narrow areas.

The Terralift soil aerator can be a tool to relieve soil compaction in certain soils. It is, however, not a universal tool for solving soil problems around trees, because the effect greatly depends on the soil type. The actual problem must be identified in the first place and the right tool chosen for the conditions. In a friction-soil (sandy soil) positive soil physical changes can be obtained but in a cohesion-soil (clay soil) the benefit is more questionable. There was a significant increase in macropores, pore volume and swell in certain horizons and this may be an indication that Terralift can create cracks and fissures that can be positive some years after treatment.

Acknowledgments

This project was financed by the Swedish Council for Building Research. I wish to thank Susanne Hansson, Magnus Nilsson and Cecilia Nylander for their assistance with the laborious field work.

Literature Cited

1. Andersson, S. 1953. Markfysikaliska undersökningar i odlad jord. II. Om markens permeabilitet (in Swedish). Grundförbättring 6: 28-45.
2. Andersson, S. 1969. Markfysikaliska undersökningar i odlad jord. XVIII. Om en ny och enkel evaporimeter (in Swedish). Grundförbättring 22: 59-66.
3. Andersson, S. och Wiklert, P. (1972). Markfysikaliska undersökningar i odlad jord. XXIII. Om de vattenhållande egenskaperna hos svenska jordarter (in Swedish). Grundförbättring 25: 53-143.
4. Bertrand, R. (1965). Rate of water intake in the field. In C.A. Black (ed.), Methods of Soil Analysis, Part 1. Physical and Mineralogical Properties Including Statistics of Measurement and Sampling. Agronomy 9 (1st ed.): 202-207. Am. Soc. of Agronomy, Madison, Wisconsin.
5. Dixon, W. J. 1965. Extraneous values. In C.A. Black (ed.), Methods of Soil Analysis, Part 1. Physical a nd Mineralogical Properties Including Statistics of Measurement and Sampling. Agronomy 9 (1st ed.): 43-49. Am . Soc. of Agronomy, Madison, Wisconsin.
6. Edling, A. P. G. 1986. Soil air. Volume and Gas Exchange Mechanisms. Sveriges Lantbruksuniversitet. Inst för markvetenskap. Avd för lantbrukets hydroteknik, Rapport 151.
7. El-Araby, A., Z. El-Haddad and M. El-Ansary. 1987. Subsoiling in some heavy clay soils of Egypt. Soil and Tillage Research 9: 207-216.
8. Harris, R. W. 1983. Arboriculture, care of trees, shrubs and vines in the landscape. Prentice-Hall Inc, London. 688 pp.

9. Hoeks, J. 1972. Effect of leaking natural gas on soil and vegetation in urban areas. Agric. Res. Rep. 778. Wageningen.
10. Hopkins, R. M. and Patrick Jr., W. H. 1969. Combined effect of oxygen content and soil compaction on root penetration. Soil Science 108: 408-413.
11. Håkansson, I. 1976. Elva försök med alvluckring och djupplöjning i syd- och västsverige 1964-1975 (in Swedish). Lantbrukshögskolan. Rapporter från jordbearbetningen Nr 42.
12. Jackson, M. B. and Drew, M. C. 1984. Effects of flooding on growth and metabolism of herbaceous plants. In T. T. Kozlowski (Ed), Flooding and Plant Growth. Academic Press Inc, New York.
13. Kozlowski, T. T. 1985. Soil aeration, flooding and tree growth. J. Arboric. 11: 85-96.
14. Lemon, E. R. and Wiegand, C. L. 1962. Soil Aeration and Plant Root Relations. II. Root Respiration. Agron. J. 54: 171-175.
15. Olsen, H. J. 1990. Construction of an Electronic Penetrometer for Use in the Field. Computers and Electronics in Agriculture 5: 65-75.
16. Richards, D. and Cockroft, B. 1974. Soil physical properties and root concentrations in an irrigated apple orchard . Austral. J. Exptl. Agric. And Husb. 14: 103-107.
17. Richards, L. A. 1948. Porous plate apparatus for measuring moisture retention and transmission by soils. Soil Sci. 66:105-110.
18. Rolf, K. 1986. Packning och packningsskador i urban miljö (in swedish). Stad och Land, Nr 50.
19. Rolf, K. 1991. Soil improvement and increased growth response from subsoil cultivation. J. Arboric. 7: 200-205.
20. Rolf, K. 1992. Soil physical effects of pneumatic subsoil loosening using a Terralift soil aerator. J. Arboric. 5: 235-240.
21. Ruark, G. A., Mader, D. L. and Tattar, T. A. 1982. The influence of soil compaction and aeration on the root growth and vigour of trees - A literature review. Part 1. Arboric. J. 6: 251-265.
22. Smiley, E. T., Watson, G. W., Fraedrich, B. R. and Booth, C. B. 1990. Evaluation of soil aeraion equipment. J. Arboric. 16: 118-123.
23. Stolzy, L. H. and Sojka, R. E. 1984. Effects of Flooding on Plant Disease. In T.T. Kozlowski (Ed), Flooding and Plant Growth. Academic Press Inc, New York.
24. Yelenoski, G. 1963. Soil aeration and tree growth. Int. Shade Tree Conf. Proc. 39: 16-25.
25. Yelenoski, G. 1964. Tolerance of trees to deficiencies of soil aeration. Int. Shade Tree Conf. Proc. 40: 127-147.
26. Youngs, E. G. 1987. Estimating hydraulic conductivity values from ring infiltrometer measurements. J. Soil Science 38: 623-632.

Personal Communications. Håkansson, Inge. Agr. Dr., Professor in soil physics at The Swedish University of Agricultural Sciences, Department of Soil Science,Uppsala, Sweden. Stamen, Ted. Urban Horticulture Advisor, University of California, USA.

Long Term, Light-Weight Aggregate Performance as Soil Amendments

James C. Patterson and Christine J. Bates

Compaction of soil as a result of visitor and equipment use of a site is a serious impediment to plant health. It can significantly alter the ability of plants to extend roots through the soil for structural support, uptake of essential nutrients, and water. This research presents results from a 22-year investigation involving several light-weight aggregates and their ability to resist compaction (lower the bulk density) resulting in a more porous soil system. At the end of the 22-year study period at the Hains Point site (silt loam soil), the bulk density of light-weight aggregate treated plots was significantly lower at 1.30 Mg/m^{-3} when compared to the control of 1.47 Mg/m^{-3}. The total pore space for the same study and treatments was significantly higher, 51.03%, compared to the control 44.48%. The same trends were realized on a fine sandy loam soil over a 21-year study period. Aggregates were incorporated at the onset of the study with only normal mowing operations occurring during the testing period while public use of the sites continued.

Soils of recreational sites are continually subjected to human impact, and thus suffer from the adverse effects of compaction. From site to site, the level of impact will vary, but the need for preventative measures remains. An impacted site such as the National Mall in Washington, D.C., is continually subjected to unrestricted use and remains a desired location for large public gatherings. In 1991, there were 2100 permitted events on the Mall, drawing participants of from several individuals to a million or more per event. Total visitation has been estimated conservatively at 13 million visitors annually. Soils on the Mall have been created by man over several hundred years, and as such, have suffered from severe compaction throughout this relatively short formation period, further complicating the soil situation. The depth of these "man made" soils ranges between 3 and 5 m. An early study of the Mall soils indicated that for three deep soil profiles, the bulk density of the surface horizon ranged between 1.96 and 2.22 Mg/m^{-3} while the total pore space for this horizon was between 13.3 and 17.2 percent (5). Similar conditions were encountered to depths of 120 cm. A more detailed study of 100 soil pits on the Mall indicated that the mean surface density (0 cm) was 1.61 Mg/m^{-3} and the subsurface (30 cm) was 1.74 Mg/m^{-3}. The mean surface pore space was 36% and the subsurface was 33% respectively (7).

A study by Veihmeyer and Hendrickson suggested that a bulk density in excess of

---

James C. Patterson and Christine J. Bates are with the National Park Service, 1100 Ohio Drive, S.W., Washington, DC 20242

1.55 Mg/m^{-3} would restrict root elongation for loamy soils (10). Vepraskas suggests that densities of 1.58 to 1.66 Mg/m^{-3} are restrictive for root extension of tobacco in sandy clay loams and sands respectively (9). The soils on the National Mall are loamy textured and thus susceptible to compaction (7). Duffy and Ralston have indicated for loblolly pine, that densities in excess of 1.45 Mg/m^{-3} can predict plant failure (2).

Clearly, soils on trails, athletic fields, playgrounds, and other areas similar to the National Mall are exposed to severe compaction. Some sites have been continually subjected to use with no maintenance efforts to curtail the adverse effects of compaction. Therefore, a long term study to evaluate soil incorporation of amendments (light-weight aggregates) to resist compaction was initiated. The study was designed and located at two sites where it was known that continuous human impact would occur and maintenance activities would be limited to mowing. Intermediate work has been published indicating short term benefits were realized over a period of 4 to 7 years with corresponding improvement of the soil conditions being realized (5). In 1992 and 1993, the research sites were revisited for sampling after 22 and 21 years respectively.

Other factors that complicate the soil conditions encountered at heavily used sites are: their method of formation by humans, the use of the site, and location (proximity to industrial activities, etc.). The soil profile is often extensively altered, thus a normal horizonation sequence no longer exists, and the existing diagnostic 'horizons' may in fact include or have been replaced by contaminates. A common feature of these soils is the existence of widely differing soil horizons and their inherent soil physical and chemical makeup. Significant physical boundaries exist between these horizons, and these are termed lithologic discontinuities. These horizons or layers of soil with widely contrasting characteristics complicate the soil physical conditions, especially the air and water movement. The soils' resemblance to its natural counterpart, therefore may be coincidental. For these reasons, detailed site investigations of existing soil conditions are strongly recommended prior to developing a site that has been exposed to human impact.

Materials and Methods

In 1966, experimentation with sintered fly ash (SFA), a manufactured byproduct of fly ash from coal-fired power plants, was used effectively to reduce soil compaction (3). The site used for this study, however, was compacted from construction but isolated from public use; thus there was a desire for the new sites to be exposed to continuous impact. The product is created from the raw fly ash dust, a particulate collected electrostatically from the smoke stacks of coal fired power plants. The raw ash can be pelletized using water followed by the sintering process which fires the material to temperatures of approximately 2000°F resulting in production of a large aggregate. These aggregates are size reduced and screened to produce the uniform material used in the studies. Advantages of this product are its angular structure, the rigid nature of the product, the internal pore space within individual aggregates (about 70%), and that it is chemically inert. Therefore, the methodology for earlier work was to incorporate the aggregate into soil in various amounts (14, 25 and 33% by volume) allowing each individual aggregate to rest one upon another providing larger macro voids between adjacent individual aggregates (3,4). An analogy would be a glass container filled with marbles, where the voids created between the marbles can be easily noted. Voids are protected from destruction, due to the ability of individual particles to resist collapse, thus providing physical support between particles.

Other available aggregates with properties similar to sintered fly ash were pursued, and expanded slate was selected. This product is heat treated in a similar fashion to SFA, is angular in nature, rigid to resist breakdown, contains internal porosity (about 70%), and is chemically inert.

The third product was coarse construction sand. It is often mentioned as a choice for soil amendment projects, but in many cases, the sand is a mixed grade containing abundant medium and fine sand particles. When mixed grade products such as this are used, they may produce a very dense final mix with the soil.

In 1971 work was begun to advance the investigation of light-weight aggregates to combat soil compaction. Two sites were selected having differing soil textures, but similar impact. The sites were Hains Point (silt loam soil), located in a playground with high visitor use, and Fort Hunt (sandy loam soil), located in the outfield of a softball field.

At Hains Point, plots were established using a randomized complete block statistical design with three replications. The sod was cut, removed, and rotary tilled for all plots except the undisturbed control. A second control was rotary tilled and sod replaced. The remaining plots received the amendments which were homogeneously tilled into the soil. The amendments used were sintered fly ash, expanded slate, coarse construction sand, and an organic additive of digested sewage sludge. Coarse construction sand was selected as an amendment because it has been believed to be an effective soil amendment in the past. One half of the plots were treated with 20 and 33% by volume of amendment and soil incorporated to a depth of 20 cm with a rotary tiller. The second half received both the amendments plus the addition of sewage sludge.

The Fort Hunt site (sandy loam soil) was implemented using the same statistical design and implementation technique as Hains Point. The only alteration in design was the elimination of the coarse sand amendment, because of the naturally sandy texture of the soil.

Analyses conducted during the study have included bulk density and calculations of total pore space and infiltration analysis. Recently root density samples have been taken to determine the root distribution characteristics within the amended soil (1). A micromorphological technique to allow visual observation of the soil pore space distribution and quantification of the pore to mineral ratio, to help define the effects of the soil amendments are planned as future study subjects. This technique may provide an effective tool to visualize and quantify the in situ effects of the aggregates and the void to mineral ratios.

The data have been analyzed using the Statistical Analysis System, version 6.08 (8).

Results and Discussion

Hains Point

Bulk density (BD) analyses have long been used as an "index of soil compaction" (6) and as an indicator of the overall physical conditions of a soil. Therefore, this research planned for tracking of the BD over the term of the project. The initial bulk density for the site was 1.51 Mg/m^{-3}. The data in Table 1 indicate that over the length of the research project, the density for the expanded slate and sintered fly ash plots were significantly lower than both rotary tilled and undisturbed control plots. As expected, the coarse sand treatment caused a significant increase in the BD and a tighter, less porous final soil mix. Upon field examination of the plot treatments 22 years after treatment, the soil structure of the undisturbed control was massive and difficult to fracture with the hands. The sand treated plots were even more difficult to fracture. Root penetration and earthworm activity as observed in the field were much less extensive in the sand treated and controls than in the light-weight aggregate treatments. The slate and fly ash treated soils were easily fractured by hand in the field, root penetration abundant, and earthworm activity was observed to be extensive, while no actual measurements were taken.

The total pore space (TPS) was calculated using a particle density (PD) of 2.65 Mg/m^{-3} and the following formula:

$$TPS = BD/PD \times 100\%$$

since traditional PD analyses call for sieving of soil materials prior to analysis, and since this is a destructive procedure, it would have eliminated the aggregates from the analysis. The TPS data for Hains Point (Table 2) indicate that for both the expanded slate and fly ash the TPS are significantly different from one another and each contains significantly more pore space than either the control or the coarse sand treatments.

The infiltration analyses for these plots were completed twice during the tenure of the study and indicated a general trend of higher infiltration than the controls but these data were not replicatable because all plots could not be treated at the same time. Also, weather conditions between field analyses caused difficulty with comparisons between and among treatments.

Root densities for the Hains Point study are not yet completed.

Table 1. Bulk Density Summary by Amendment for Hains Point—Summary for the Entire Project

Amendment Type	Bulk Density Mg/m^{-3}	Duncan's Group*
Coarse Sand	1.49	A
Undisturbed Control	1.47	B
Rotary Tilled Control	1.46	C
Sintered Fly Ash	1.34	D
Expanded Slate	1.30	E

*Duncan's multiple range test, SAS 6.08. Means with the same letter are not significantly different at the 0.01 probability level.

Table 2. Total Pore Space Distribution by Amendment for Hains Point—Summary for the Entire Project

Amendment Type	Pore Space* (mean %)	Duncan's Group**
Expanded Slate	51.03	A
Sintered Fly Ash	49.37	B
Rotary Tilled Control	44.90	C
Undisturbed Control	44.48	D
Coarse Sand	43.94	E

*Total pore space was calculated using 2.65 Mg/m^{-3} as the particle density.

**Duncan's multiple range test, SAS 6.08. Means with the same letter are not significantly different at the 0.01 probability level

Initial micromorphologic analysis has shown that the amendments create larger macro voids and root penetration between and through these voids occurs. Figure 1 is a photomicrograph of the aggregate - pore space arrangement. Along the lower edge of the aggregate in the upper left corner of the micrograph, a root is penetrating the pore space between the aggregates. This space, represented by the white voids, provides evidence that pores remain in existence in this soil 22 years after treatment with the light-weight aggregate amendments. As noted, this site has been exposed continuously to compaction with no treatment beyond the initial incorporation of the amendments. It is hypothesized that "bridging" exists between individual aggregates and this may provide the ability for these soils to retain internal porosity. This feature is yet to be confirmed.

Fort Hunt

BD analyses followed the same trends as for the Hains Point site with the expanded slate and sintered fly ash significantly different from one another and each significantly less dense than either control (Table 3). The BD of the sewage sludge amendments was between that of the two controls. The BD for both the slate and fly ash treatments is somewhat below the ideal BD of 1.33 Mg/m^{-3}. The BD of the slate treatment is 1.19 Mg/m^{-3} and for fly ash 1.23 Mg/m^{-3}. This low BD is likely due to the light-weight nature of these two amendments.

The TPS relationships for the Fort Hunt study appear in Table 4. The slate and fly ash treatments exhibited significantly higher TPS than either control or the sewage sludge treatments, and there was a slight reduction in TPS because of the rotary tillage

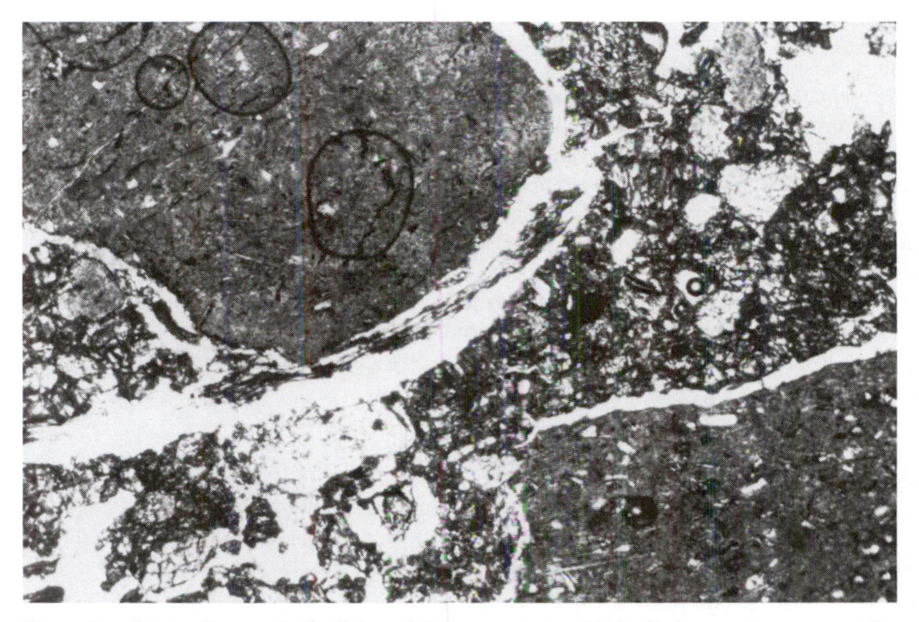

Figure 1 — Photomicrograph of a light-weight aggregate amended soil. Aggregates appear at the upper left and lower right, the pore space appears white, and the soil mattrix lies between the aggregates. A root is apparent in the pore space beneath the aggregate in the upper left of the micrograph.

Table 3. Bulk Density Summary by Amendment for Fort Hunt—Summary for the Entire Project

Amendment Type	Bulk Density Mg/m^{-3}	Duncan's Group*
Rotary Tilled Control	1.35	A
Sewage Sludge	1.32	B
Undisturbed Control	1.32	C
Sintered Fly Ash	1.23	D
Expanded Slate	1.19	E

*Duncan's multiple range test, SAS 6.08. Means with the same letter are not significantly different at the 0.01 probability level.

of the sandy loam soil. In some cases tillage can destroy soil structure.

Difficulties occurred when attempting to complete large numbers of infiltration analyses in a timely fashion, because of changeable daily weather conditions and the lengthy time involved with this analytical technique. As a result, both between treatment and within treatment errors were realized.

The root density analyses mirrored those of the Hains Point plots with more rooting and earthworm activity being observed in the slate and ash treatments than the controls.

Micromorphologic analyses are being conducted to determine if the same trends occur as were noted with the other aggregate-treated plots.

The question of the ideal percentage of amendment to be included in any soil mix is open to conjecture. A key question is, "How much amendment is necessary to efficiently resist compaction?". To address this issue, Table 5 combines the data from both sites. It should be noted that both the 20 and 33% by volume incorporation of the amendments are significantly different from one another as well as significantly different from the control treatments. Whether the 33% by volume or a higher volume of amendment incorporation is necessary may be determined by the proposed use for the site. If heavy impact and use is anticipated, then it seems prudent to incorporate the higher percentage of amendment. If the use is expected to be intermittent and generally low, then the lower percentage may be the most effective mixture. The cost of these amendments is approximately $50.00 per metric ton, and cost may dictate how much will be used. It is necessary, however, to fully assess the expected use, and thoroughly incorporate the appropriate amendment into the soil during the site implementation phase in order to provide a soil condition to resist compaction and provide a suitable recreation surface.

Conclusions

Expanded slate and sintered fly ash have shown an ability to withstand the forces of soil compaction over a 22-year study period. Slate and fly ash significantly reduced bulk density with a corresponding increase in total porosity. Earthworm and root penetration in the amended plots was improved and this appears to support the existence of improved soil physical conditions, especially the soil structure.

Initial micromorphologic analyses of amended soils appear to support the hypothesis that macro pores occur between adjacent aggregates and that roots were able to grow into these pores. This is an area of future research focus.

Table 4. Total Pore Space Distribution by Amendment for Fort Hunt—Summary for the Entire Project

Amendment Type	Total Pore Space* (mean %)	Duncan's Group**
Expanded Slate	55.08	A
Sintered Fly Ash	53.65	B
Undisturbed Control	50.28	C
Sewage Sludge	50.25	D
Rotary Tilled Control	49.21	E

*Total pore space was calculated using 2.65 Mg/m^{-3} as the particle density.
**Duncan's multiple range test, SAS 6.08. Means with the same letter are not significantly different at the 0.01 probability level.

Table 5. Percent of Amendment Incorporated with the Soil—Summary for the Entire Project

Amendment (% by volume)	Pore Space* (mean %)	Duncan's Group**
33	48.85	A
20	47.33	B
0	44.69	C

*Total pore space was calculated using 2.65 Mg/m^{-3} as the particle density.
**Duncan's multiple Range test, SAS 6.08. Means with the same letter are not significantly different at the 0.01 probability level.

It is advisable to consider properly amending the soil in the design and planning phase, since it seems virtually impossible to adequately amend a site soil once plant material has been established.

Literature Cited

1. Black, C. A. (ed). 1965. Methods of Soil Analysis. Part 1. American Society of Agronomy, Inc. Agronomy 9: pp. 371-377.
2. Duffy, P. D. and D. C. McClurkin. 1974. Difficult eroded planting sites in North Mississippi evaluated by discriminant analysis. Soil Sci. Soc. Am. Proceedings 38:676-678.
3. Patterson, J. C. 1969. Sintered fly ash as a soil modifier. M.S. Thesis, Dept. of Agronomy, West Virginia University, Morgantown, WV. 114 pp.
4. Patterson, J. C., and P. R. Henderlong. 1970. Turfgrass soil modification with sintered fly ash. Proceedings of the First International Turfgrass Research Conference, Harrogate, England. pp. 161-171.
5. Patterson, J. C. 1976. Soil compaction and its effects upon urban vegetation. In Better Trees for Metropolitan Landscapes Symposium Proceedings. U. S. Forest Service General Technical Report NE-22, pp. 91-102.
6. Pierre, W.H., D. Kirkham, J. Pesek, and R. Shaw. (ed). 1966. In. Plant Environment and Efficient Water Use. American Society of Agronomy and Soil Science Society of America. p. 111.

7. Short, J. R. 1983. Characterization and classification of highly man-influenced soils of the mall in Washington, D.C. M.S. Thesis, University of Maryland, College Park, MD. 126 pp.
8. Statistical Analysis System, 1990. SAS Circle, Box 8000, Cary, NC. 27512-8000.
9. Vepraskas, M. J. 1988. Bulk density values diagnostic of restricted root growth in coarse -textures soils. Soil Sci. Soc. Am. J. 52:1117-1121.
10. Veihmeyer, F. J., and A. H. Hendrickson. 1948. Soil density and root penetration. Soil Sci. 65:487-493.

The Effect of Trenching on Growth and Overall Plant Health of Selected Species of Shade Trees

Fredric D. Miller, Jr.

New fiberoptic telephone lines were installed in trenches throughout the campus of the University of Illinois at Urbana-Champaign in the spring of 1987. The trenches were in close proximity to tree trunks. Annual growth and mortality data were taken on *Celtis occidentalis*, *Liquidambar styraciflua*, *Acer saccharum*, and *Gleditsia triacanthos* through 1991. Only 7 of 98 trees died during the trial period. Trenching distances of 0.5 to 3.3 m. did not predispose the trees to readily evident disease or insect infestations. Only on *Celtis* was there statistically different growth between trenched and control trees for all growing seasons.

Shade trees contribute value and attractiveness to urban landscapes and city streets. However, disruptive activities such as street widening, trenching for utilities, grade changes or material storage (including soil) may result in damage to tree root systems and increased stress for trees. According to Harris (3) and Zimmermann and Brown (6) the majority of tree roots are concentrated in the top 1 m (3 ft) of the soil and if growing in an open space may spread at least as far as 2-3 times the height of the tree. Disturbing the soil around the tree may cause serious damage to the root system resulting in the possible decline or death of the tree.

Trenching around trees for the installation of utilities (ie. telephone lines) can be very disruptive to root systems particularly when done within the dripline of the tree. Not only is the soil disturbed, but roots may be severed and exposed to the air resulting in possible additional injury of root tissues. Morell (4) found that trenching due to water main installation resulted in tree mortality rates of 25-44% with 100% mortality in certain cases approximately 12 years after the trenching activity had occurred. Current recommendations as proposed by BSI (2) and Watson (5) suggest a minimum distance for trenching along one side of the tree of 0.15 m (0.5 ft) for each 2.5 cm (1 in) dbh. Harris (3) and ASCA (1) recommend 0.3 m (1 ft) for each 2.5 cm (1 in) dbh. For augering along one side of a tree, Morrell (4) recommends that trenching be stopped approximately 0.3 m (1 ft) from the tree for each 2.5 cm (1 in) dbh.

In the spring 1987, telephone lines were installed throughout the University of Illinois at Urbana-Champaign campus and as a result a number of trenches were excavated in close proximity to the trunks of various species of shade trees along the campus and

---

Fredric D. Miller, Jr. is with the University of Illinois Cooperative Extension, 6438 Joliet Road, Countryside, IL 60525.

city parkways. In order to examine the long term effect of trenching on certain tree species a study was begun in the spring of 1987 with the following objectives: 1) to examine the long term effect of trenching on the growth and plant health of selected tree species particularly as it relates to the proximity of trenching to the trunk and annual incremental growth, and 2) to monitor the long term effect of trenching on these shade trees and their predisposition to attack by insects and woody plant pathogens.

Materials and Methods

In the spring of 1987, prior to the initiation of plant growth, sites were selected on the University of Illinois at Urbana-Champaign campus that contained blocks of at least 9-10 trees of the same species in which trenching had occurred as a result of telephone line installation. The tree species selected for study included hackberry, *Celtis occidentalis* (dbh = 23.4 - 26.2 cm), sweetgum, *Liquidambar styraciflua* (dbh = 28.2 - 28.4 cm), sugar maple, *Acer saccharum* (dbh = 29.0 - 29.5 cm), and honeylocust, *Gleditsia triacanthos* (dbh = 31.8 - 47.7 cm). Non-trenched trees of the same species and comparable size class, and located as near to the trenched trees as possible served as controls. During the 1987 growing season, the hackberry control trees experienced an unknown malady resulting in partial defoliation and chlorosis of the foliage. As a result, a new set of control trees within one block of the trenched trees was selected as replacement. Therefore, there were no growth data for the hackberry control trees for the 1987 growing season. Trenches were 0.6 to 0.9 m (2 - 3 ft) deep and 25.4 to 30.5 cm (10 - 12 in) wide (Figs. 6&7)

In order to evaluate the effect of trenching on these particular tree species, diameter at breast height (dbh) measurements (nearest 0.01 mm) were taken for both the trenched and control tree groups. In the early spring of 1987, dbh measurements were taken in order to establish a baseline dbh value. Subsequent dbh measurements were taken each year in either the late winter or early spring following each growing season to determine annual incremental growth rate. The study was terminated following the 1991 growing season. A 5 cm (2 in) diameter dot of black spray paint was applied to the trunks of all trees to insure consistency in dbh measurements from season to season. Periodically, during each of the five growing seasons, observations were made as to any insect or disease problems and overall tree health. Linear measurements were also taken from the trench to the trunk of the affected trees.

Weather records for the 1987-1991 growing seasons (March - October) were obtained from the Illinois State Water Survey Research Center, Champaign, Illinois. Data were analyzed using the Solo Statistical System, Version 2.0 (BMDP Statistical Software).

Results and Discussion

Overall Growth Effects

A summary of overall growth for trenched and control trees for the 1987-1991 growing seasons is presented in Figs. 1 - 5. Of the four tree species observed in this study, only hackberry showed a consistent statistically significant difference in growth between the trenched and control trees for all growing seasons (Fig. 1). Statistically significant differences in growth between trenched and control trees were observed for sugar maple (1987) (Fig. 3), sweetgum (1991) (Fig. 2), and honeylocust (Site #2) (1988,1990) (Fig. 5),

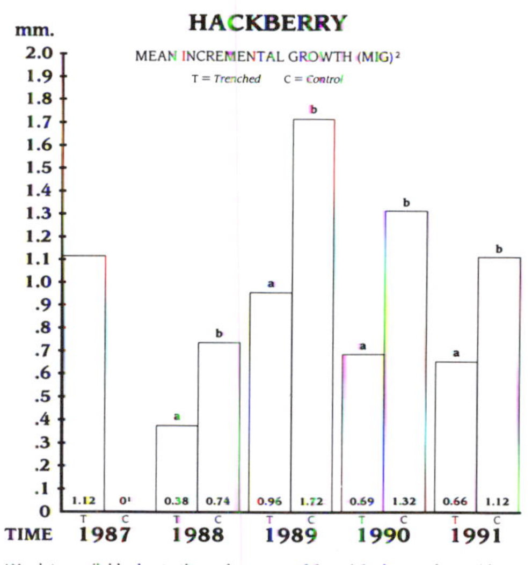

Figure 1 — Mean incremental growth of trenched and control hackberry trees.

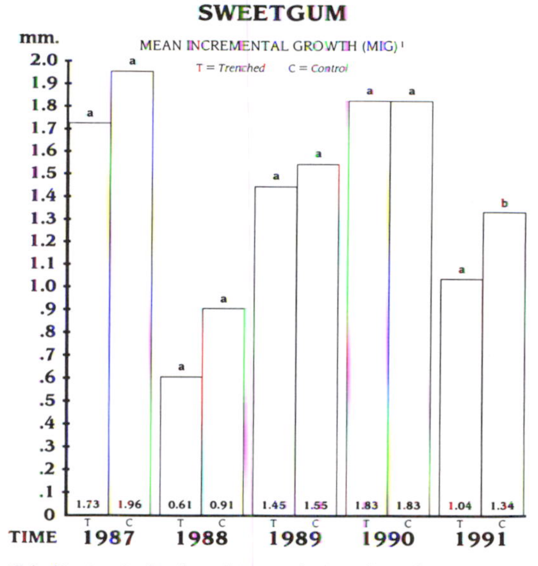

Figure 2 — Mean incremental growth of trenched and control sweetgum trees.

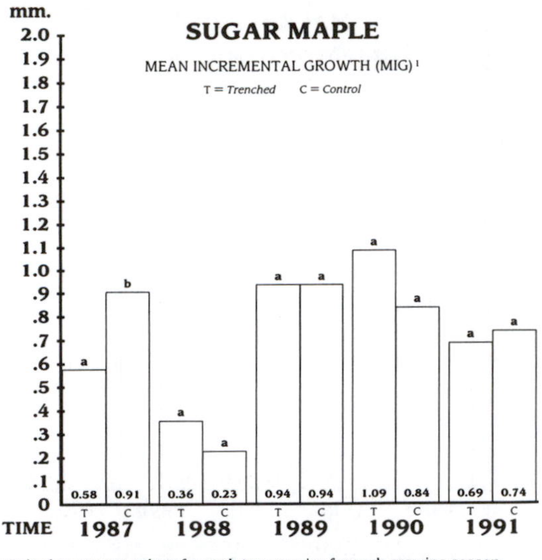

[1] Paired treatment values for each tree species for each growing season followed by the same letter are not significantly different. ($p<0.05$; Two-Tailed T-test).

Figure 3 — Mean incremental growth of trenched and control sugar maple trees.

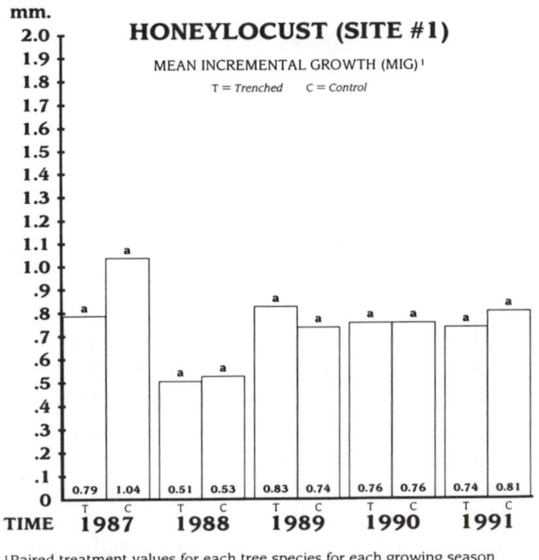

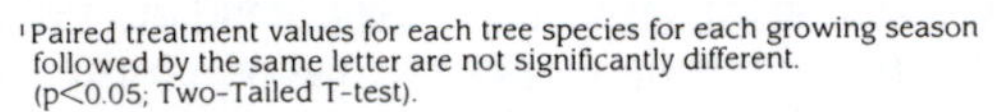
[1] Paired treatment values for each tree species for each growing season followed by the same letter are not significantly different. ($p<0.05$; Two-Tailed T-test).

Figure 4 — Mean incremental growth of trenched and control honeylocust (site #1) trees.

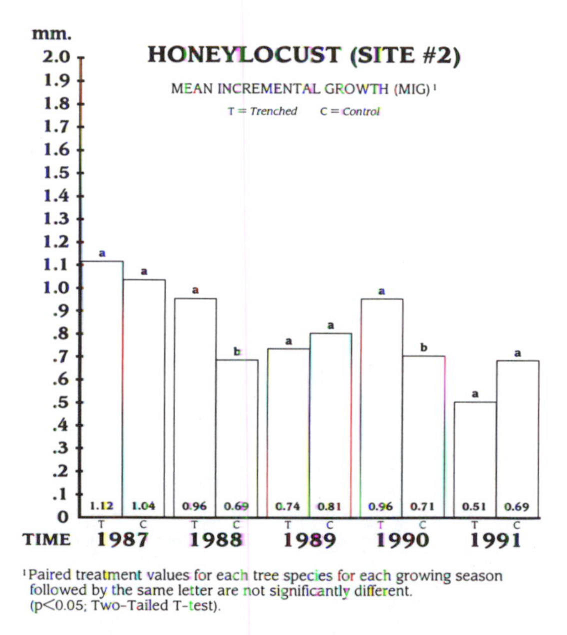

Figure 5 — Mean incremental growth of trenched and control honeylocust (site #2) trees.

but were not consistent throughout the five year study. Differences in growth were also observed with honeylocust (Site #1), but they were not significantly different statistically (Fig. 4).

Growth measurements taken in the fall of 1987, following the trenching activity the previous spring revealed no significant difference in mean incremental growth (MIG) rates between trenched and control trees except for sugar maple (Fig. 3). Mean maximum temperatures were near normal for June-August and rainfall totalled 17.5 cm (6.9 in) above normal. The month of July was the sixth wettest recorded. However, following the 1988 growing season, all trenched and control trees experienced considerable reductions in MIG of 44% and 50%, respectively (Figs. 1 - 5). The extensive drought that occurred throughout central Illinois and the north central United States may have been one of many factors responsible. The study area experienced rainfall levels 15 cm (5.9 in) below normal for March through August with June being the driest month ever recorded. Overall, 1988 was the seventh driest year on record. In conjunction with the drought-like conditions, mean maximum temperatures were approximately 2.3°C (5°F) above normal with June - August having the third highest mean maximum temperature of 32.0°C (89.6°F). More normal rainfall amounts and normal temperatures occurred during the 1989 and 1990 growing seasons. Trenched and control trees exibited MIG increases of 39% and 44%, respectively (Figs. 1 - 5). In contrast to 1988, adequate rainfall and moderate temperatures may have contributed to the observed increase in MIG. As in 1988, drought-like conditions prevailed again in 1991. Rainfall levels were 17.8 cm (7 in) below normal for June - August resulting in the third driest June in central Illinois since 1888 and the ninth driest summer on record. Temperatures during this same period were 4°C (8.7°F) above normal with 46 consecutive days with temperatures

Figure 6 — Trenching along one side of hackberry, *Celtis occidentalis* trees for installation of telephone lines on the University of Illinois, Urbana-Champaign campus, spring, 1987. (Photo credit: James Smith, University of Illinois at Urbana-Champaign.

equal to or greater than 32.0°C (90°F). Trenched and control trees grew 27% and 10% less, respectively as compared to the 1990 growing season (Figs. 1 - 5).

Trenching Distance and Tree Growth

Results from this study seem to indicate that reduction in growth is affected by the proximity of trenching activity. The greater the distance of trenching from the trunk, the less reduction in growth. Hackberry with trenching within 0.5 m (1.5 ft) of the tree trunk had an overall mean growth reduction of 45% over five consecutive growing seasons (Fig. 1). In contrast, sweetgum, sugar maple, and honeylocust trees, with trenching at greater distances, had growth reductions of less than 15% (Figs. 2 - 4). Honeylocust trees (site #2), trenched 3.3 m (11 ft) from the trunk, had a reduction in growth of only 7% (Fig. 5).

Recommendations by the British Standards Institute (2) (1989) and Watson (5) suggest a distance of 0.15 m (0.5 ft) for each 2.5 cm (1 in) dbh when trenching along one side of a tree. Harris (3) citing work by Kimmel (1978, personal communication) and ASCA (1) specify for utility installation a minimum distance for trenches of 0.3 m (1 ft) per 2.5 cm (1 in) dbh of trunk. Trenching on the U of I campus occurred much closer than the recommended BSI and Watson distance for hackberry, sugar maple, and honeylocust (site #1) trees. Trenching for the honeylocust (site #2) trees occurred slightly beyond the minimum distance recommended by BSI and Watson. The data from this study support these minimum trenching distance recommendations since less than 15% growth reductions occurred in three of the four species examined (Figs. 2 - 5). In addi-

Figure 7 — Soil compaction following trenching along one side of hackberry, Celtis occidentalis trees for installation of telephone lines on the University of Illinois, Urbana-Champaign campus, spring, 1987. (Photo credit: James Smith, University of Illinois at Urbana-Champaign).

tion, field observations revealed no apparent differences in the shape of the crown of trenched trees as compared to control trees.

Tree Mortality

Only seven trees out of a total of 98 trees (trenched and controls) died during the course of the study for an overall mortality rate of 7.1%; five of the trees were honeylocust. Two of the trees (one hackberry and one honeylocust) were removed due to ice storm damage in the spring of 1990 and two honeylocust trees were removed due to cankers. The remaining three trees were removed for causes unknown. Six of the seven trees that died had been trenched. With the exception of the cankers on honeylocust, which may have developed due to stress, no major insect or disease problems were observed on any of the tree species throughout the five year study.

Conclusions

Results from this study indicate that as the trenching distance from the trunk increased, the impact on potential tree growth decreased. Trenching did not necessarily predispose trenched trees in this study to catastrophic insect and disease infestations in spite of two growing seasons with drought-like conditions nor was there any extensive mortality of any of the tree species after five years of observations. The trees adapted and survived. However, in no way do these findings minimize the importance of stress factors such as trenching on woody plant material and their vulnerability to insect and disease infestations. It is fully recognized that other variables such as tree species, genetic differences, soil conditions, and environmental affects may all play a role in annual incremental growth. Further study is needed to determine the relationship between these factors, and overall tree growth and plant health.

Acknowledgments

The author wishes to express sincere thanks to Mr. James Smith, Horticulturist with the University of Illinois Operations and Maintenance Division for assistance in locating the study trees and photographs of the trenching activities. Special thanks is extended to K. von der Heide-Spravka of Downers Grove, Illinois for assistance in data entry and statistical analysis.

Literature Cited

1. American Society of Consulting Arborists. 1989. Protecting trees during construction: Answers to frequently asked questions for builders and property owners. Wheat Ridge, Colorado. 1 pp.
2. British Standard Institute. 1989. British Standard Guide for: Trees in Relation to Construction. British Standard Institute. Publication #10285. 26pp. (In press).
3. Harris, R.W., 1983. Arboriculture: Care of Trees, Shrubs, and Vines in the Landscape. Prentice-Hall, Inc. Englewood Cliffs, New Jersey. 688 pp.
4. Morell, J.D. 1984. Parkway tree augering specifications. J. Arboric. 10 (5): 129 - 132.
5. Watson, G. 1990. Preventing construction damage to trees. The Morton Arboretum information leaflet #36. Lisle, Illinois. 1 pp.
6. Zimmermann, M.H. and C.L. Brown. 1971. Trees: Structure and Function. New York: Springer-Verlag.

Aspects of the Damage to Asphalt Road Pavings Caused by Tree Roots

Jitze Kopinga

Studies into the mechanisms by which tree roots (especially of poplar) cause damage to asphalt road pavings have revealed that a constant high moisture content of the soil directly underneath the pavement surface can be considered as a major soil factor in attracting the trees root to develop there. This paper presents an overview of several field studies on the nature and background of the damage in The Netherlands and the results of some experiments that have been set up to develop methods to control or prevent damage by tree roots. Of these methods, thus far the use of rubble as road bed filling appears to prevent root development for a period of at least five years.

Almost everywhere in The Netherlands, both in the urban and the rural areas, the roots of tree plantings along roads cause damage to the road surfaces. Generally, only the minor roads are involved, such as the so-called land reclamation roads, service roads along motorways and, of course, the many cycle tracks that are present in The Netherlands. The problems occur most frequently on roads with thin asphalt surfaces such as cycle tracks and pedestrian walkways. Root damage always has the financial disadvantage of road repair costs, but it also results in unsafe traffic situations because of the dangerously irregular road surface and the tendency of bikers and motorcycle drivers to avoid such cycle tracks and take the main road or sidewalk.

For many road managers the phenomenon of root damage has always been a mystery. What are those long roots searching for underneath the tarmac, and why do only a few of them pass under the road so quickly in order to form an intensive root system again once they have reached the other side? Does such a root know that there are other environments suitable for rooting at a distance of sometimes more than ten meters from its origin, or is there something else going on?

At the request of the Road and Hydraulic Engineering Division of the Ministry of Transport and Public Works of The Netherlands, in 1984 the Institute of Forest and Nature Research started a long term research project to reveal the backgrounds of this 'mystery' and to develop usable and practical control methods to prevent root damage. Design, construction and maintenance of the roads as well as the trees standing alongside them were all to be considered.

This paper presents a short overview of a number of results and findings of this project thus far that can be of practical use. The aim is not to cover in detail the various experiments that have been set up within the framework of this project or its evaluation. These are described elsewhere (e.g. 4) or will be published in the course of time.

Jitze Kopinga is with the Instituut voor Bos- en Natuuronderzoek (IBN-DLO). Bosrandweg 20, P.O. Box 23, NL 6700 AA Wageningen, The Netherlands.

Investigation of the Problems

In 1984 and 1985 the first field surveys were made, and an enquiry was held among the road managers of the Ministry of Traffic and Public Works in order to determine where and when damage occurs. The following aspects were taken into account:

- the character of damage and the rooting pattern underneath the road surface;
- the tree species and cultivars that may cause damage;
- construction and nature of the road;
- age and development of the tree plantings;
- planting distance and distance between trees and road surface;
- quality of the tree site;
- rates of root development underneath the road surface.

The Character of Damage and the Rooting Pattern

The typical damage to road surfaces is a pattern of crack formation that, as a rule, runs more or less transverse on the road or centrifugal from the stem base of the tree (Plate 1). Further opening of the cracks reveals the presence of long roots with diameters from 2 to 2.5 cm (sometimes even less). It is remarkable that the rate at which the roots branch is very low and there is also little formation of small lateral roots.

Intensive rooting occurs again on the other side of the road, once it is reached by the tree roots. Also remarkable is that the roots are only present close to the underside of the road surface (Plate 2). Only occasionally are roots found deeper in the road bed.

Sometimes the damage also consists of warty bumps ('cauliflowers') that bulge upwards. This has been observed most often with poplar. A closer examination reveals that the road surface is pushed up by the formation of so-called root suckers. Generally these are formed by the roots on spots where they are mechanically damaged or when the still living root is severed from the trunk.

Tree Species and Cultivars That May Cause Damage

In Table 1 tree species that have been found to cause 'typical' damage are presented.

Plate 1 — Example of the typical pattern of damage to a tarmac pavement by tree roots, in this case by poplar.

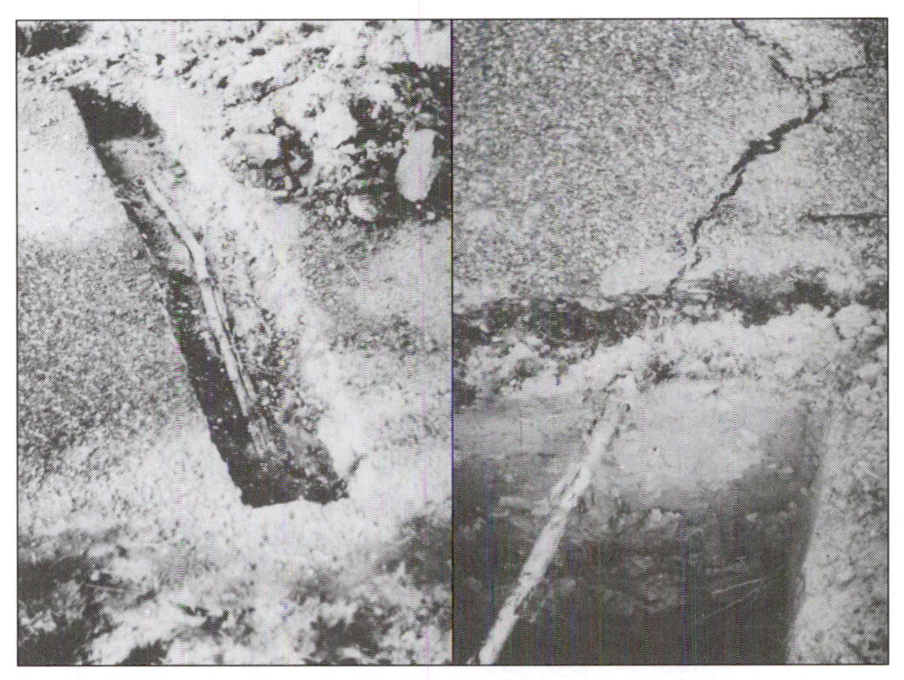

Plate 2 — The damaging roots are localized directly below the underside of the pavement.

The species are ranked according to the number of observed cases.

It should be noted that this ranking does not take into account the frequency in which the named species is used as a roadside tree, and therefore does not necessarily reflect the comparative behaviour of the tree species themselves. It should also be noted that all the mentioned species belong to the group that is known in forestry as typical pioneer species. In their responses to the questionnaire, the road managers occasionally also mentioned other species such as penduculate oak, beech, ash and elm. But a closer field examination at the indicated locations showed little of the characteristic pattern of damage from the long, comparatively thin roots. In most cases damage was only observed on the edge of the road surface as a result of the thickening of the roots at the stem base of trees that were standing very close to the road.

Construction and Nature of the Road Surface

In most situations, the roads were constructed on a sand bed approximately 30 cm thick that was mechanically compacted and then paved over with one or two layers of a warm asphalt and gravel mixture. These were than rolled even and sometimes also covered with a thin layer of fine crushed rubble. The total thickness of the road surface varied from 8 to 12 cm and the width from approximately 2 m (cycle tracks) to more than 3.5 m (land reclamation roads, parallel roads or service roads).

Damage by tree roots was also observed on cycle tracks paved with tiles of 30 x 30 x 5 or 7 cm. This category of road is only a minor part of the total of cycle tracks, because of the relatively high costs of construction and service, and was therefore not taken into further consideration.

Table 1. Species of trees and shrubs that are commonly found to cause root damage to road surfaces in the Netherlands

Common Name	Botanical Name
Poplar	Populus (all common species)
White willow	Salix alba
Common birch	Betula pendula
Black locust	Robinia pseudoacacia
Common pine	Pinus sylvestris
Buckthorn	Hippophae rhamnoides*
Silver maple	Acer saccharinum**

*Especially in recreation sites along the sea coast
**Especially in urban areas (the species is used only on a small scale in rural areas)

Age and Development of the Tree Plantings

Generally, damage appears to occur from the moment the tree or shrub plantings are half grown. However, there are some exceptions, such as Euramerican poplar where damage was observed when plantings were only 5 to 7 years old.

Planting Distance and Distance Between Planting and Road

Damage was observed both along forest type plantings and along row plantings of trees. There was no distinct correlation between the distance of the trees to the roads and the frequency in which damage occurs, more or less regardless of the age of the planting and road surface. As extreme examples, there were situations in which 20-year-old poplars standing approximately 1 m from the road still showed no damage, while in other situations damage had already occurred on road surfaces that were more than 7 m from the stem base of the same row of poplars (corresponding to a distance of approx. 1 m outside the crown projection of the trees).

Quality of the Tree Site

Damage was observed in both relatively high humus-rich clay soils as well as in sandy clay (or loamy) soils and poor sandy soils from drift-sand, fluvial depositions or heathland clearings. There was no indication that the presence or absence of ground water and the explorability of it by the tree roots was of any influence on the occurrence of damage. In fact, damage was observed with more or less the same frequency on both the higher dry soils and the lower wet soils.

In a few cases, in clay-peat soils, a pattern of damage was observed in which crack formation was not transverse but more or less parallel to the direction of the road. Because it was evident that tree roots were not directly causing these cracks, no further attention was paid to this phenomenon. A more recent study of these locations has shown that this phenomenon only occurs on shrinkable clay soils where differences in the swelling and shrinking of the soil occur as a result of an irregular pattern of water attraction out of the soil by the roots of the roadside trees (1).

The question of whether a restricted rootable soil volume could be the stimulus for the tree roots to escape underneath the tarmac could not be answered positively. There appeared to be no distinct correlation between underground growing space (as calculated from width of the roadside soil, planting distance and rooting depth) and the rate in which damage occurred. Even in situations where rooting space was largely sufficient, according to some recent Dutch assumptions and directives (3,5,7), damage could frequently be observed.

There are important exceptions, however. In a number of typical urban situations, trees are planted in much too narrow planting holes in soils that cannot be penetrated by tree roots because of too high soil compaction. Thus the roots can only escape to better surroundings via the underside of the road pavings. In such conditions even species other than the 'aggressive' tree species, mentioned in Table 1, cause root damage to the pavements.

During the research period, soil samples were occasionally taken to determine the chemical soil fertility of both the roadside soil and the bedding of the roads. In all cases, it appeared that the sand bed consisted of poor material (low contents of organic matter, potassium, magnesium and directly available phosphorus) and that the quality of the roadside soil, with a few exceptions, satisfied the minimal demands for acceptable tree growth (2).

The Rate of Root Development Underneath the Road Surface

Generally, roots are able to transverse the width of a cycle track within 1 to 2 years. This is illustrated by the results of a study of a 2.2 m wide road near an approx. 20-year-old row of poplars which was carried out as part of this project. Over some distance, on both sides and close to the edge of the road, twelve roots were cut and studied for age and diameter. The average age of the roots on the roadside facing the trees was 5.8 ± 1.8 years and the root pieces had an average diameter of 32.0 ± 7.8 mm. The age of the roots on the other side was 5.1 ± 1.6 years and their average diameter was 29.7 ± 10.0 mm. The difference in age of the roots on both sides of the road was 0 years for five of the roots, 1 year for five of the roots and 2 years for the other two. The observed damage was already two years old and this again illustrates that within a few years after the construction of a cycle track along a fully grown poplar planting damage by "just" thumb-thick roots may occur.

Study into Some Physical and Chemical Factors in the Soil

Because the results of the field investigations gave no clear answer to the question of how roots could be "forced" or "invited" to develop underneath the road surface, it was decided to perform a more detailed study of some field situations to examine additional growing site factors and possible fluctuations in them during the growing season.

The examined factors were: penetration resistance, moisture supply or humidity, the temperature and oxygen content of the air in the soil, both beside and below the surface of cycle tracks on two locations in the Netherlands. Apart from this field study, a supplementary container trial was set up at the trial field of the IBN to examine if asphalt itself, as a chemical substrate, might have any stimulating or inhibiting effect on the growth and development of tree roots.

Penetration Resistance of the Soil

Penetration resistance of the soil was determined by using a penetrograph (manufacturer Eijkelkamp, type 'Stiboka', cone surface: 1 cm^2 with a top angle of 60 degrees). Measurements were taken in the spring, while installing the measuring devices. An illustration of some of the results is presented in Figure 1.

Given that the upper limit of root penetration in soils is about 3.0 MPa, it is clear that the degree to which the bed of the road is compacted is far too high to allow root penetration, and that the verge can be penetrated fairly easily. Therefore, it can be concluded that once they have left the verge, roots can only develop and grow in close contact with the underside of the road. The reason for this is the presence of an interface between two different materials (with different rates of expansion when they are heated and cooled) which permanently or occasionally results in some small space for root penetration.

Judging from Graph B in Figure 1, it could be concluded that the upper zone of the bed is not compacted, but this is a method-linked measuring artefact. Depending on the condition and nature of soils, the meter will always indicate a too low degree of resistance up to a depth of 5 to 10 cm because soil material can be pushed to the side and up.

Fluctuation of the Soil Moisture Content

The soil moisture content is determined both manometrically, with tensiometer cups (for the low values of suction), and electrically with moisture blocks (for the higher values). An illustration of the fluctuation of the soil suction in the course of the growing season of 1986 is presented in Fig. 2. It should be noted that 1986 was a dry year in which little rain fell during the growing season. The graphs clearly indicate that the degree of humidity of the soil below the road surface is constantly high, while the soil beside the road gradually dries out (depending on the rate of rainfall). The different rates of soil suction between the verge and the soil underneath the asphalt therefore could be one of the explanations for root development underneath road surfaces.

Supplementary research has revealed that the quantity of water underneath the asphalt is at 'field capacity'. This means that there will not be a mass flow of water and that this zone becomes dry rather quickly when the water is absorbed by the tree roots. Any replenishment of water, for example by the condensation of water vapour on the underside of the pavement, will still not be sufficient to cover the water demand of the tree. This was illustrated in one of the field trials in which roots caused a continuously high value of the readings of moisture blocks, once they had grown close to them.

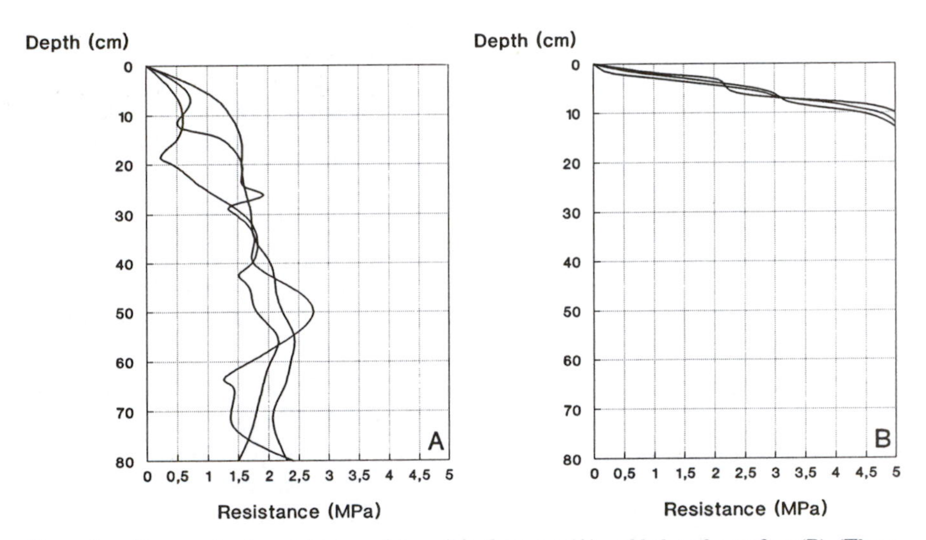

Figure 1 — The penetration resistance of the soil in the verge (A) and below the surface (B). (Three measurements per graph.)

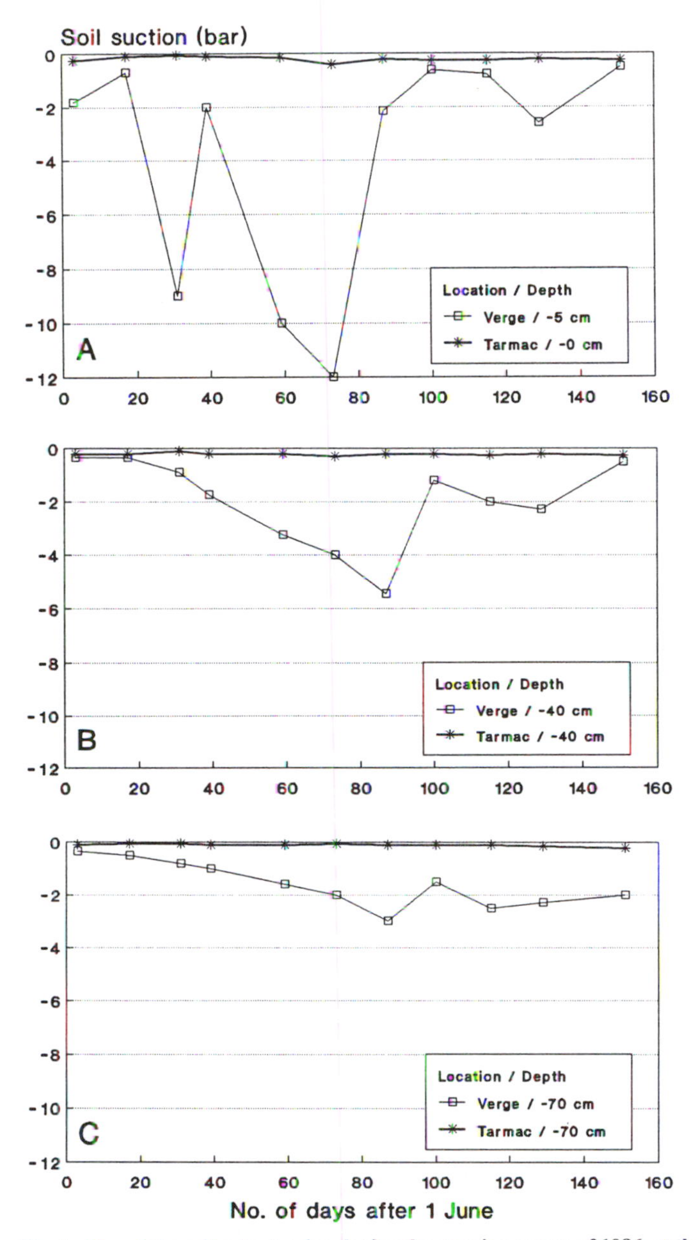

Figure 2 — Fluctuation of the soilwater tension during the growing season of 1986, at three depths (A,B,C) below and besides the pavement. Each point represents the average readings of six separate sensors.

Soil Temperature

An illustration of the fluctuation of daily soil temperature below and beside the road surface is presented in Fig. 3. It appears that during the summer the average temperature below the road surface is higher than at comparable depths in the roadside soil. To what extent this could influence the rate of root growth, however, cannot be determined as no diurnal fluctuations have been measured and it is not known how much the asphalt cools down during the night. Also, the average magnitude of the differences between the road-side soil and the road surface during periods of high activity of root development (early spring and mid-autumn) is unknown.

Oxygen Content of the Soil Air

Both in the roadside soil and below the pavement, down to a depth of 70 cm, the oxygen content of the soil air appeared to fluctuate almost continuously between 15 and 19% during the growing season. This oxygen content is remarkably higher than the minimum of 10%, below which stagnation of root activity can be expected. The fact that even underneath the air-impermeable tarmac the soil maintains a sufficient oxygen level can be ascribed to a sufficient sideward transport of soil air as a result of the relatively high air diffusion rates of the sand in the road bed, which is between 0.01 and 0.05 cm^2/sec (6). From these findings it can be concluded that soil oxygen deficiency is not a restricting factor when small roads are concerned and that therefore the absence of roots deeper in the bedding must have other causes (e.g. high penetration resistance).

The Effect of Asphalt as a Chemical Substance on the Development of Tree Roots

In the course of the study, the question arose as to what extent the asphalt itself, as a chemical substance, could have an effect on the occurrence of damage by tree roots. Although this supposition was not supported by field studies that indicated that root damage could be observed in both asphalt pavements and pavements of concrete tiles, the question was considered intriguing enough to justify a closer study. Because data about the chemical composition and water-solubility of the various asphalt compounds provided insufficient information to answer the question, some aspects were also studied in a container trial with Euramerican poplar (cv. Robusta). Rooted cuttings of the trees were planted in mixtures of grated tarmac (derived from a damage location) and sand (poor sand or potting soil/sand-mixture) at various mixing rates. The results of the trial indicated not positive, but rather a negative, effect of any of the concentrations of asphalt mixtures on the rate of root growth. Although the circumstances of the experiment can not be compared with normal field situations, and therefore the results of the trial must be considered with some reserve, on the basis of these results, combined with existing data on the water-solubility of asphalt compounds, it may be stated that if the tarmac should have any effect on the development of damage by roots, the effect is not likely to be substantial.

The "Model" of Root Damage

Based on the aforementioned considerations and study results, a model for the development of roots underneath road surfaces was designed. Attracted by the comparatively high degree of humidity of the soil underneath the road surface, the roots penetrate the bedding of the road. Root development is restricted to the interface between the sand and the pavement, because elsewhere in the bedding the development of roots is prevented by the compaction of the soil. Once the roots have arrived underneath the pavemant, the small quantity of water present is absorbed quickly by the roots, which causes a gradient of soil humidity that stimulates a rapid apical growth of the roots. This, in

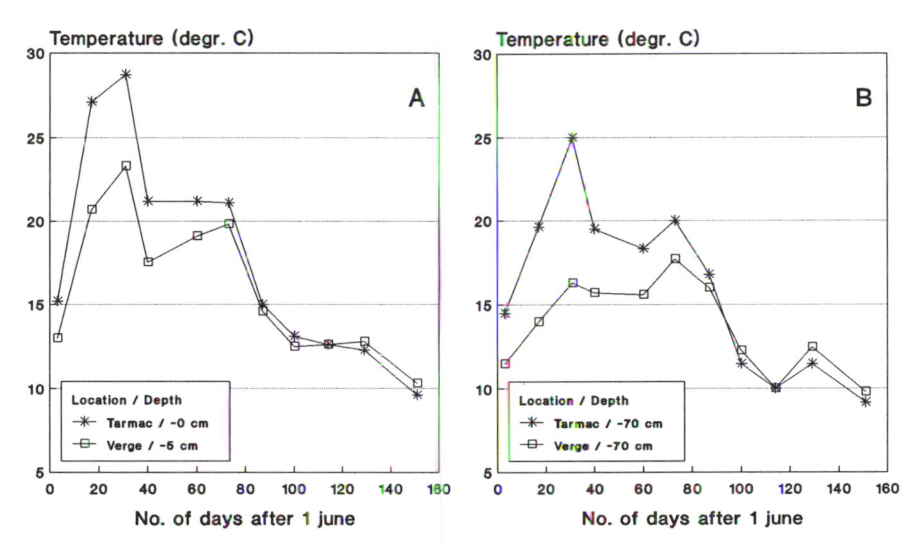

Figure 3 — Fluctuation of daily temperature (measured between 11 a.m. and 3 p.m.) during the growing season of 1986 at two depths (A,B) below and beside the pavement. Each point represents the average readings of six separate sensors.

combination with the low level of soil fertility, is probalbly the reason that roots underneath the pavement form few branches and fine lateral roots. Once the roots have reached the soil on the other side of the road, they explore in a normal rooting pattern. This induces an increase in the diameter of the roots underneath the pavement, which after some period of time, depending on factors such as thickness and elasticity of the pavement, results in the formation of cracks.

Of course this general model cannot answer all of the questions connected with the problem. It does not explain, for example, why some tree species such as poplar, willow and birch are especially likely to cause damage while other species such as linden and ash hardly ever do so. Presumably, a number of other qualities of the trees or the "behaviour" of their root systems are responsible for this. To develop a more comprehensive model, more would have to be known, for example, about each species' rooting capacity in compacted soils or differences in their responses to a moisture-gradient in the soil (8).

Methods to Control or Prevent Damage by Tree Roots

The most effective way to prevent damage by tree roots, of course, is to abandon the idea of constructing roads near existing tree plantings or planting trees near existing roads. This is not a realistic option in The Netherlands, however, because of considerations of landscape planning. Therefore other solutions have to be found.

The use of trees with less "aggressive" root systems could be a good solution, but this is also not always desirable from a landscape planning view point. Moreover, it is not opportune in existing situations. In these cases a number of technical solutions must provide some relief.

Some technical methods have been tested in a long-term study at the trial field of the IBN on a cycle track construction of about 200 m length. The pavement consists of

'Road Tape' (a few millimeters thick, bitumen-impregnated material, comparable with the so called roofing leather that is widely used in house-construction) and it was covered with fine-grade rock rubble. The trees along the cycle track were three-year-old Euramerican poplars (*Populus* x *euramericana* 'Robusta') that were planted at two meter intervals and one meter away from the road. To check whether the soil site of the test construction were similar to those of the already studied field situations, moisture and temperature sensors were installed at three different depths, both underneath and beside the pavement, and at intervals of ten meters. The methods tested included:

- the use of rubble as road bed filling;
- the application of a root growth inhibiting herbicide Preventol-B;
- the use of so called anti-root barriers. The effects of the use of anti-root barriers is also being studied in a series of excavations of field situations in which various materials were already applied on a practical scale.

Rubble as Road Bed Filling

The material that was used consisted of debris of bricks from some of the stone mills in the Betuwe, that was ground to course fragments. It was applied in a layer about 30 cm thick. One year after construction of the trial (which was in the beginning of 1987) it appeared that no roots had developed directly underneath the pavement in these sections of the road. Under all other treatments root development could be observed. Root development remained absent in the rubble till February 1992, when the entire trial was excavated and studied for the first time. Root development was indeed present, however, on spots where sand had been washed in between the fragments as a result of rainfall.

A possible explanation for the absence of root development, in spite of the continuously high level of humidity in the zone directly underneath the pavement, is that the mechanical resistance of the rubble and probably also the irregular structure of the interface between road bed and pavement (the rubble was not sprinkled over with sand before covering) forms too large an obstruction for the tree roots. Also, the space between the fragments might be too large for the roots to bridge.

Root Herbicides

The root herbicide Preventol-B was chosen for the trial. Mixing this agent with the sand of the road bed, however, was considered not opportune, because of the political and environmental concerns. Therefore the choice was made to use the already existing Preventol-B impregnated roofing material as a layer between the sand and the pavement. It was thought that the sideward contact of the roots with the impregnated material could provide a sufficient inhibition of root development to prevent the roots from growing under the pavement. A second consideration was that the presence of the substance in only the tarmac might be tolerable from an environmental point of view, because asphalt itself is regarded as an unnatural material and therefore the discharge, recycling and eventual dispersal of the material in the environment can be controlled better than for road sand.

One year after the trial was set up there seemed to be some inhibition of root growth by the application mentioned here, as judged by visual inspection. But after 5 years, there was no distinct difference in comparison to the control sections of the trial.

From this result it may be concluded that the application of a root herbicide in only the pavement material is obviously insufficient to prevent root growth. However this assumption should be made with some reserve, because the test involved only one of the many possible methods of application, one of the existing compounds, and at only one concentration.

Screens of Anti-root Barriers

The anti-root barrier used in the test was Terram-1000, a so-called non-woven cloth of polymer fibers, also known as Geotextile. The material is about 0.7 mm thick and it was laid below the road bed over lengths of 10 metres of the road, with an upturned edge reaching above the mowing field on the side facing the trees. After the first growing season, root development could be observed underneath the pavement as a result of sideward root penetration through the material. After five years there appeared to be a slight but not significant difference between the weight of roots in these sections compared with the control sections. The average fresh weight of the roots per tree was 1887 ± 975 grams (for 12 trees) to 2153 ± 810 grams for the control trees.Another trial studied whether or not a thicker material could possibly prevent root penetration. Bottomless containers were used, which were placed on various types of Geotextile, with thicknesses ranging from 0.4 to 1.4 mm, that was rolled out over the soil of the trial field. The containers were filled with the same soil material that was underneath the geotextile and planted with rooted cuttings of ‘Robusta’ poplar (see Plate 3).After the first year, root penetration could be observed on all types of Geotextile. However, there appeared to be distinct differences in the number and diameter of roots that had penetrated the textiles (Fig. 3).

A practical implication of the aforementioned findings is that the tested materials are not suitable as an anti-root screen in situations where damage by roots must be prevented with a 100% certainty. It could be expected that materials other than the non-woven materials, such as the plastic sheets that are used in agriculture, might prevent root penetration. However, the use of such materials in the road bed are not advocated for civil-technical reasons. It is assumed that it would stagnate the draining of the sand bed, which would result in frost damage to the pavement during the winter. However,

Plate 3 — Configuration of the container trial to test root penetration of various types of Terram Geotextile. A: spring 1987, B: Autumn 1988.

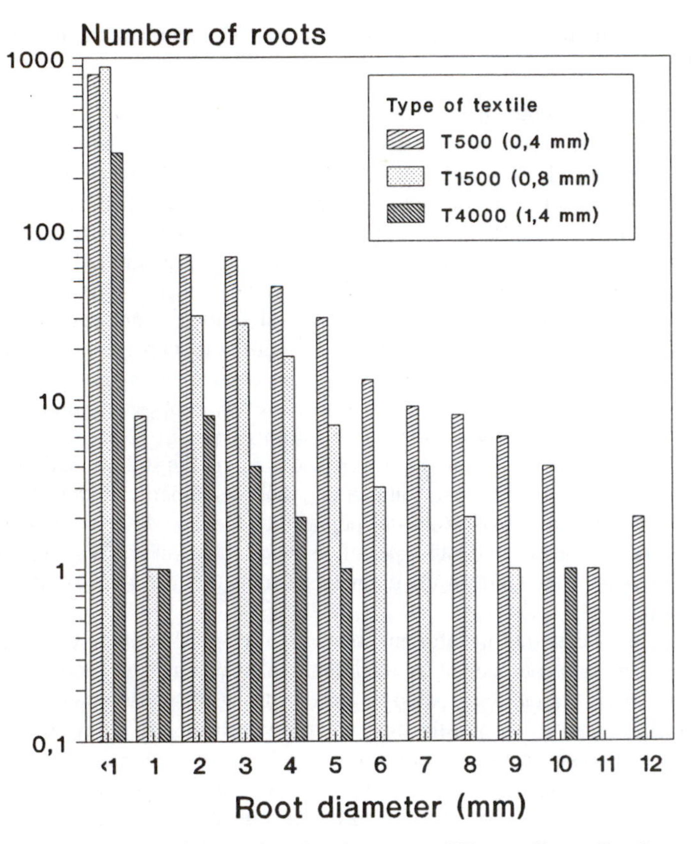

Figure 4 — Difference in root penetration of various types of Terram Geotextile after one growing season. The figure represents the total number of roots per diameter-class of 12 trees.

the materials are used in practice as a vertical root barrier along the edge of roads.

Vertical Anti-root Screens

In 1988 an investigation was carried out at 16 locations where it was known that some years before vertical screens had been installed along the pavements of cycle tracks and pedestrian walkways. Various materials were used for this, both water-impermeable plastic sheets and water-permeable non-woven materials as well as sheets of woven plastic ribbon. The screens were excavated on both sides of the tracks over a length of several meters and the pattern of root development was observed. The first above-ground inspection made it clear that in none of the 16 cases had the screens been effective in preventing damage by roots. It appeared that the roots had grown over or below the screens. Where sheets of plastic ribbon were used, the roots had even grown straight through. The growing over by roots could be ascribed to mechanical damage (by mowing equipment) or weathering (by sunlight) of the parts of the screens that reached above ground. Remarkably, where foils of non-woven materials were present no root penetration was observed (in contrast to the results in the previous trial).

From these findings, the conclusion can be drawn that one must not expect too much from root screens in situations where the rooting depth is deeper than the underside of the screen and where above ground damage of the screen cannot be excluded.

Summary

Although knowledge of the mechanisms of damage to pavements by tree roots has increased substantially in the seven years since the research project began, and some practical directives have found a sound basis, not all the causes of root damage nor the possibilities for its prevention have been revealed. It is to be expected that more insight in the near future will be gained by further studies and the research that is still in progress. For example, experiments to test how the road bed can be made unattractive for tree roots in a relatively "environmentally friendly" way. Nevertheless, at the moment it is known that root barriers of plastic screens or similar materials will only be sufficiently effective when they permanently reach above and below the rootable zone of the soil. Before a road manager decides to install such a barrier, he first has to make sure that this condition is met in practice. At locations where the average lowest ground water level is not too deep and also forms the limit of rooting depth, the screen might be embedded in cement mortar under the ground water level so that some kind of mechanical barrier is created for roots growing downwards directly alongside the screen (and again upwards on the other side). There is strong evidence that a road bed of clean, medium-ground rubble does not allow development of roots underneath the pavement, provided the bed is not sprinkled over or impurified with sand. One also may consider the use of other materials besides rubble, such as course lava slag or sinters from the ore-melting industry (provided the latter have a sufficiently low content of heavy metals). Less definitive, but still encouraging are the indications that the thicker type of non-woven plastic fiber sheets strongly inhibit the thickening of fine roots after they have penetrated the material, because the roots will be strangulated by the cloth around the penetration points. Large scale application has to be regarded with some reserve, however, because the long term effects of this method are not yet known.

Literature Cited

1. Bakker, J.W., P. Cornelissen & Th. H. van de Putten. 1982. Droogteschade aan wegen bij bermbeplanting. Wegenbouwkundige werkdagen deel II, 20 - 21 May 1992, C.R.O.W., Ede: 693-704.
2. Bemestingsadviesbasis voor Stedelijk Groen inclusief stadsbomen en sportvelden. Rapport nr. 604, De Dorschkamp, Instituut voor Bosbouw en Groenbeheer, Wageningen, 1990, 60p.
3. Kopinga, J. 1985. Research on street tree planting practices in The Netherlands. Proc. 5th conference of METRIA: Selecting and Preparing Sites for Urban Trees, Pennsylvania State University, 23 - 24 May, 1985, 72-84.
4. Kopinga, J., G.J. Jansen & C. Das. 1989. De achtergonden van het ontstaan van schade door boomwortels van met name populier aan wegverhardingen en de toetsing van enige methoden om schade te beperken of te voorkomen. Dienst Weg- en Waterbouwkunde, Delft, 1989, 112p.
5. Kopinga, J. 1991. The effects of restricted volumes of soil on the growth and development of street trees. J. Arboric. 17 (3): 57-63.

6. Kopinga J. & J.W. Bakker. 1992. Zeer-open-asfaltebeton als alternatief voor tegelverhardingen op de doorwortelde bodemzone bij (straat)bomen. Dorschkamprapport nr. 689, Instituut voor Bos- en Natuuronderzoek, Wageningen 1992, 25p.
7. OBIS. 1988. Bomen in straatprofielen: groeiplaatsberekening, voorbeeldbladen. Vereniging Nederlandse Gemeenten (VNG), 's-Gravenhage, 63p.
8. Wiersum, L.K. 1984. Verslag van het onderzoek aan wortels van een aantal boomsoorten, mede in verband met onze samenwerking met de OBIS. Nota nr. 130. Instituut voor Bodemvruchtbaarheid, Haren (Gr.), 17 p.

Root Barriers for Controlling Damage to Sidewalks

Philip A. Barker

Several types of root barriers tested in field experiments in northern California inhibited development of shallow tree roots. The need for an improved barrier design to prevent development of potentially harmful circling roots inside a barrier was indicated. Roots grew over the tops of barriers that had been accidentally covered with even thin layers of soil, cancelling out barrier benefits.

Radial growth of tree roots often warps and cracks overlying sidewalks, creating "lips" by uneven displacement of adjoining sections of sidewalks. Such damage impedes safe usage of the sidewalks, resulting in pedestrian accidents, commonly classified as "trip and fall," and claims for damages by injured victims. Root damage to sidewalks is economically important in U.S. cities (1,7,8) and elsewhere, including Mexico (4) and England (9).

Contrary to common viewpoint, sidewalks apparently promote rather than deter development of shallow tree roots (2). A concrete sidewalk prevents soil moisture loss by either evaporation or transpiration and blocks percolation of rainwater into the soil as well. Moreover, a sidewalk warms rapidly and radiates heat to the soil beneath. It likewise cools more rapidly than underlying soil. Consequently, moisture from the soil condenses on the underside of the sidewalk, only to evaporate back into the soil whenever heat buildup of the sidewalk again outpaces that of the soil (5).

The timeframe between the planting of street trees and their damage to adjacent sidewalks undoubtedly is substantially shorter for trees that develop shallow rather than exceptionally deep roots. This is because the forces generated by radial growth of deep roots, compared to those of shallow roots, should dissipate throughout a larger volume of soil before impacting any overlying sidewalks.

In cooperation with Solano Community College, the USDA Forest Service operates a 5-acre site in northern California as the Solano Urban Forestry Research Area (SUFRA) for conducting field research on control of rooting depth of trees (Fig. 1). Being about 10 miles north of the delta region and mouth of the Sacramento River and a similar distance inland from the north end of San Francisco Bay across a range of low mountains, the climate at this field facility has a maritime influence. Winter temperatures rarely go below freezing, and rainfall averages about 15 inches annually, occurring primarily from October through April.

The deep, well-drained alluvial soil, with a pH range of 6.5-7.5 and electrical con-

---

Philip A Barker is a Research Horticulturist with Urban Forestry Research, U.S. Forest Service, P.O. Box 245, Berkeley, CA 94701.

Figure 1 — American sweetgum trees (*Liquidambar styraciflua*) in an experiment at the Solano Urban Forestry Research Area. Planted 3 years earlier as container-grown, 6-foot-tall stock, the trees are ready for excavation of their roots to a depth of about 1 foot in an area within a 3-foot radius from trunk center. The exposed roots then will be excised and their dry weights per tree determined as the basis for assessing treatment effects.

ductivity, representing soluble salts, of 300-500 micromhos/cm, is a dark brown, generally silty clay loam, without mottling, typical of soils in the Class I Yolo Series (6).

Trees planted with different treatments to the root systems are grown 3 or more years with turfgrass cover and sprinkler irrigation. Afterwards, the roots of each tree are excavated and harvested to an approximate depth of 1 foot within a 3-foot radius from trunk center (Fig. 2) to determine their dry weight.

Root barriers made of either rigid plastic materials or rhizotoxic fabric, namely Biobarrier™, and installed as planting hole liners to a 12-inch depth have effectively inhibited development of shallow roots (Barker, unpublished data). However, circling roots induced by each type of barrier tested (Fig. 3) suggest the need for a barrier designed to retard such root growth. With this in mind, an experiment was installed in spring 1993 to determine within 2 to 3 years if various makes of root barriers with internal ribs inhibit circling roots and how much this effect differs among the barriers.

Whatever the barrier type, the top edge must be installed and kept permanently above

Figure 2 — Roots are excavated manually by first trenching around the perimeter of the intended excavation area, then dragging soil from the excavation area into the trench and alternately filling and emptying the trench until the pit has been excavated to the desired depth.

ground, because, where buried under either soil or mulch, we have found that roots overgrew the barrier and benefits were lost (Fig. 4). Also, a barrier should not be installed so deep that development of a tree's critical anchor roots is compromised. Research with circular barriers at the Solano Urban Forestry Research Area has shown, as previously stated, that barriers only 12 inches deep effectively inhibit development of shallow roots; however, the effect may differ among soils and tree species. Until future research is done to identify optimum barrier depth, a prudent approach would favor installing a barrier too shallow rather than too deep.

Root barriers that fit snugly around a rootball like a sausage casing, as opposed to barriers installed as planting hole liners, likewise have effectively inhibited development of shallow roots (3). An experiment to be installed within 6 months will test the effectiveness of a casing type of root barrier specially designed for simplicity of use. This root

Figure 3 — Barrier-induced circling roots were a common problem in an experiment that tested the effectiveness of three 12-inch deep root barriers in inhibiting development of shallow roots. Circling roots developed inside rhizotoxic root barriers but were particularly abundant inside plastic barriers which lacked internal vertical ribs.

barrier is nested in the last container a tree is shifted into before it is marketed (Fig. 5). If results of this experiment are positive, availability of containerized trees with built-in root barriers may supercede the need to purchase and install trees and root barriers as separate entities.

Linear barriers are an alternative to surrounding the roots of trees with circular barriers. These are installed alongside a sidewalk or other paved surface, either during replacement of a damaged sidewalk and simultaneous pruning of offending tree roots, or when new trees are planted. Their effectiveness has yet to be tested on a quantitative basis.

In summary, research at the Solano Urban Forestry Research Area has shown that root barriers of various types can inhibit development of shallow roots. Implicit in these results, however, is the need for follow-up research that includes such treatment variables as soils, tree species, and sidewalks.

Figure 4 — Any benefits that root barriers may provide, regardless of type, were nullified where even thin layers of soil covered a barrier's top edge. Root overgrowth is no less a problem where the top edge of a root barrier is obscured by mulch, as shown in the bottom photograph. With vigorous-growing tree species, such as cottonwood, *Populus* sp., root overgrowth may occur even if the top edge of the barrier protrudes well above grade and never has been covered by mulch.

Figure 5 — Prototype of a root barrier (leaning against a whole container) for nesting in a whole container when a tree is transplanted in it. Upon planting the tree in the landscape, the rootball is extracted from the whole container, leaving the prepackaged root barrier intact.

Acknowledgment

Use of trade or firm names in this paper is for reader information and does not imply endorsement by the U.S. Department of Agriculture of any product or service.

Literature Cited

1. Barker, Philip A. 1983. Some urban trees of California: Maintenance problems and genetic improvement possibilities. In Gerhold, Henry D., ed. METRIA: 4, Proceedings of the fourth biennial conference of the Metropolitan Tree Improvement Alliance, 1983 June 20-21, Bronx, NY. University Park, PA. The Pennsylvania State University, School of Forest Resources; 1983: 47-54.
2. Barker, Philip A. 1988. Proactive strategies to mediate tree-root damage to sidewalks. Combined Proceedings, The International Plant Propagators' Society 37(1987):56-61.
3. Barker, Philip A. 1990. Tree roots and sidewalk conflicts. In Make our cities safe for trees, Proceedings of the Fourth Urban Forestry Conference, St. Louis, Missouri, October 15-19, 1989, pp. 134-136. Washington, DC. American Forestry Association.

4. Benavides Meza, Hector M. 1992. Current situation of the urban forest in Mexico City. J. Arboric. 18(1):33-36.
5. Harris, Richard W. 1992. Arboriculture—Integrated Management of Landscape Trees, Shrubs, and Vines, 2nd ed. Englewood Cliffs, NJ : Prentice Hall. 674 pp.
6. Soil Conservation Service. 1977. Soil survey of Solano County, California. USDA Soil Conservation Service, Davis, CA. 65 pp.
7. Sommer, Robert, Christina L. Cecchettini, and Hartmut Guenther. 1992. Agreement among arborists, gardeners, and landscape architects in rating street trees. J. Arboric. 18(5):252-256.
8. Wagar, J. Alan, and Philip A. Barker. 1983. Tree root damage to sidewalks and curbs. J. Arboric. 9(7):177-181.
9. Wong, T. W., J. E. G. Good, and M. P. Denne. 1988. Tree root damage to pavements and kerbs in the city of Manchester. Arboric. J. 12:17-34.

A Buyer's Technical Guide to Root Barriers

Paula J. Peper and Philip A. Barker

Various commercially available root barriers with internal vertical ribs are compared on a performance basis to augment their descriptions by manufacturers.

The frequently described conflict between tree roots and sidewalks remains one of the most pervasive problems confronting urban forest managers (2,3). Research indicates that root barriers can substantially reduce root biomass in the top foot of soil within a 3-foot radius from the tree trunk (Barker, unpublished data). Whether such results translate into less conflict between tree roots and sidewalks over time is still in question. Current manufacturers continue to design and produce new barriers and additional companies are entering the marketplace for the first time. As competition and barrier variety increase, consumers need unbiased sources of information to determine whether or not to use barriers and, if they are to be used, which barrier will best suit particular site requirements.

Numerous kinds of root barriers are commercially available for an urban tree manager who considers root barriers worth using. An early root barrier, introduced in 1976, is marketed as the Standard Deep Root Planter™ (Patent No. 4019279). Made of thick plastic, this sturdy barrier is 18 inches deep and 22 and 29 inches square at the top and bottom, respectively. Other barriers now being marketed are designed for ease of packaging and shipping. These are either flat panels for connecting together at point of use to make whatever length is desired or they are continuous barriers that are rolled up for shipping and may or may not be pre-cut to a specified length. Either type of barrier may be installed linearly along sidewalks or cylindrically as planting hole liners.

Of the numerous root barriers available, most of them have internal vertical ribs intended to direct roots downward and thus retard development of circling roots. This paper discusses characteristics of four panel type and two continuous barriers, each with internal vertical ribs. The paper is based on our observations during and following installation in May 1993 of the panel type barriers in a field experiment at the Solano Urban Forestry Research Area (SUFRA) in northern California and on our observations of the continuous barriers in actual use. Each of these barriers is briefly described, alphabetically by manufacturer, in Table 1.

Important in selecting a barrier is the material it is made of and various design features, notably the way in which either the panels or the ends of continuous barriers are connected together. Panels may be joined together with interlocking couplings, separate

---

Paula J. Peper is a Biological Technician and Philip A. Barker is a Research Horticulturist with Urban Forestry Research, U.S. Forest Service, P.O. Box 245, Berkeley, CA 94701.

Table 1. Manufacturers, materials, prices, and design features of five ribbed root barriers, listed alphabetically by manufacturer

Manufacturer	Material	Panel depth in.[1]	Quoted retail price per linear foot	Rib length	Rib height in.	Rib type and angle degrees	Watering tubes attached to barrier	Panel connecting device [2]	Other features
Bumble Bee Products, Inc.,	Polyethylene,	12	$1.70	Entire	9/16	90	No	Separate	Notched at
3260 Industry Dr.,	high-density,	24	3.25	depth	9/16	90	No	extruded	bottom for
Signal Hill, CA 90806	injection-molded			of panel				connector	tearing by
310-597-7933									maturing roots
Century Products	Polystyrene,	18 x 69	2.74	Entire	5/8	45[3]	Yes	Connector	
1401 N. Kraemer Blvd. #B	extruded,	18 x 120	2.56	depth	5/8	45[3]	Yes	glued w	
Anaheim, CA 92806	w/ultra-violet	24 x 69	3.44	of	5/8	45[3]	Yes	chloromethane	
714-632-7083	stabilizer	24 x 120	3.06	panel	5/8	45[3]	Yes	solvent	
Deep Root Partners, L.P.	Polypropylene,	12	2.18	Ends 1/4"	1/2	90	No	Separate	Antilift pads
345 Lorton Ave. # 305	injection-molded	18	3.28	from	1/2	90	No	extruded	to keep panels
Burlingame, CA 94010		24	4.10	panel	1/2	90	No	connector	from lifting
800-458-7668				bottom					once installed
Shawtown Industries, Inc.	Polystyrene,	12	2.25	Ends 1/4"	1/2	90	No	Inter-	Tight fit of
4550 Calle Alto, Unit D	extruded	18	3.50	from top	1/2	90		locking	locking device
Camarillo, CA 93010		24	4.60	and	1/2	90	No	coupling	deters panel
800-772-7668		48[4]		bottom	1/2	90	No		slippage
Vespro, Inc.[5]	Polyethylene,	12		Entire	Varies	"agonic	No	Inter-	
40 Belvedere St., Unit 2	low density,	18		depth of	1/4 to	curl,"	No	locking	
San Raphael, CA 94901	extruded	24		panel	3/4	90° arc	No	coupling	
415/459-7311									

[1] All panels 24 inches wide unless otherwise noted.
[2] Illustrated in Figure 1.
[3] Hollow triangular tubes molded to inside wall of barrier.
[4] Special order.
[5] Retooling to produce a different design.

connectors or locking strips, or by use of chemical bonding agents (Fig. 1). As indicated in Table 1, the panels may be made of polyethylene, polypropylene, or polystyrene. Among these three thermoplastics, polyethylene in the high density formulation is recognized in the plastics industry as most resilient and durable. By comparison, polypropylene is slightly harder and therefore eventually may chip or crack easier. Polystyrene, on the other hand, crystallizes readily in the presence of sunlight, in which case its durability is compromised.

Panel Barriers

We installed 46 each of the panel-type Bumble Bee, Deep Root, and Vespro barriers in two field studies at SUFRA in May 1993. Each barrier, installed as a planting hole liner, consisted of three panels connected with either separate locking strips or interlocking couplings. Two-year-old, bare root seedling trees, each approximately 6 feet tall, were planted by backfilling with unamended native soil inside the barriers after which gaps between the exterior of the barriers and the planting holes were collapsed with a shovel. When a tree was planted, the top edge of each barrier extended above grade at least 1 inch. Three months after installation, eight barriers if each kind were randomly selected and examined.

Characteristics unique to each barrier are discussed below in alphabetical order.

Bumble Bee Barrier (Patent No. 4995191)

A circular shape of this high-density polyethylene barrier was easily maintained while the trees were being planted. Of particular importance, neither the panels nor the connectors were predisposed to slipping out of alignment while the soil was being backfilled.

The 8 randomly selected trees for follow-up examination exhibited no signs of chipping or breakage of the rim above grade level. Uneven settling of panels was not apparent; however, we did observe chipping and tearing—probably due to impacts by mower wheels—of a few of the internal vertical ribs that protruded above grade level. Ribs are designed to redirect root growth downward, and any that are broken provide an opportunity for roots to grow in a circular pattern until they meet an intact rib.

In a 3-year old installation of these barriers at the University of California in Davis, some of the panel connectors had already cracked or ruptured above grade. We understand the manufacturer now makes these connectors with a more durable polystyrene.

Deep Root Barrier

A circular shape of this polypropylene barrier was easily maintained while a tree was being planted. On the other hand, because of looseness of fit of the installed locking strips, they had to be held in place and often repositioned to line up with the top edge of the panels during soil backfilling.

Three months later, one of the eight randomly selected Deep Root barriers exhibited settling of individual panels, leaving top rims only half an inch above grade level. "Anti-lift pads," which are small tabs running horizontally around the inside of these barriers to stabilize the panels and keep them from lifting once installed, may actually have a reverse effect. Gravity, along with weight of the water-saturated soil inside the barriers may be causing them to settle. Also, the top edge or rim on one of the barriers had a 4-inch long tear at ground level, obviously resulting from a mower wheel hitting it.

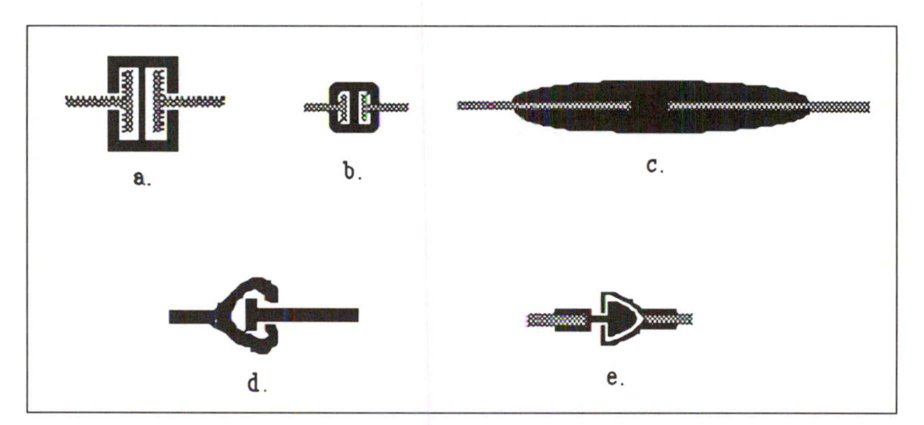

Figure 1 — A schematic of cross sections of root barrier panel connectors. Deep Root (a) and Bumble Bee (b) barriers use separate locking strips that slide over panel ends. Century (c) uses chloromethane solvent to bond connector to panels. Vespro (d) and Shawtown (e) have interlocking coupling connectors that are either extruded with the panel or bonded to the panels during manufacturing. Approximate scale.

Deep Root barriers installed approximately 3 years ago at Dan Foley Park in Vallejo, California, reveal minor chipping and cracking above grade level.

Shawtown Barrier

We have recently installed samples of this barrier at SUFRA to observe the long term effects of environmental exposure, particularly sunlight, on barriers made of polystyrene. During installation, the circular shape of the barriers was easily maintained. The interlocking coupling on the barrier is bonded to the panels at the factory and we saw no evidence of panel slippage after we assembled them.

Vespro Barrier

This barrier, among the three types of barriers used in the previously mentioned field studies at SUFRA, was most cumbersome to handle. Because it was made of low-density polyethylene, it was extremely malleable, becoming more so as temperatures increased. During installation, therefore, one person had to continuously hold it in a circular shape while another person backfilled the soil within the barrier. Even with this extra help, the installed barrier had an uneven undulating shape, which complicated the mowing and edging of turfgrass surrounding the barrier.

The connecting device of this barrier is an interlocking coupling, which is extruded as part of each panel. Despite the simplicity of this coupling and, therefore, ease of connecting the panels, separation of the panels after barrier installation posed a problem. Panels separated once when we inserted one of two tree stakes in the backfill soil inside the barrier after the tree had been planted. This required panel replacement and replanting of the tree. A close examination of the coupling revealed that it was uneven because of faulty extrusion. It was also possible to separate connected panels by pulling them apart by hand. In short, the pliancy of this barrier compromises its structural integrity.

Two of the eight Vespro barriers examined 3 months after installation had vertical tears in the 1-inch, above-grade segment, at the mold joints of the internal ribs. Individual panels on four of the barriers had settled unevenly, but still remained above grade level. There was no evidence of further separating of panel connectors; however, undulation of the cylindrical shape of the barriers had become more pronounced.

Continuous Barriers

We installed samples of continuous barriers marketed by Century and Shawtown at SUFRA in June 1993 to observe long-term effects of environmental exposure. Both barriers maintained their shape as trees were being planted.

Century Barrier

Assembly of these barriers as they were being installed required use of chloromethane solvent (methylene chloride) to bond the locking mechanism to the panels. Although easy to use, chloromethane may pose health hazards to installers. Failure to glue the connector to the barrier ends could allow separation of the barrier ends when or shortly after a tree is planted and growth of tree roots through the gap. There is no experimental evidence that the hollow triangular tubes glued onto the inside wall of this barrier will prevent circling roots, as intended. Nor is there experimental evidence that soil aeration and water application is effectively enhanced with these tubes or with larger watering tubes that are sometimes glued onto the outside wall of this barrier.

Shawtown Barrier

Observations of Shawtown continuous barriers installed in 1992 along a sidewalk in Fresno, California, revealed extensive breakage from rim tops to ground level. This was no surprise because elsewhere we have seen polystyrene barriers begin breaking and crumbling within 1 to 2 years, due to its tendency to crystallize and become brittle when exposed to sunlight.

Materials Testing

Pamphlets distributed by the manufacturers to advertise barriers provide information on the tensile, flexural, and impact resistance properties of their products (Table 2). This

Table 2. Alphabetical listing of results of engineering tests reported in barrier manufacturers' brochures.

Barrier	Material thickness, mil (1/1000 in.)	Tensil strength, (ASTM D 638), psi	Flexural properties (ASTM D 790)		Impact resistance (ASTM D 256)	
			Strength, psi	Elasticity, psi	Izod., ft.-lb.	Gardner/ Rockwell, in.-lb
Bumble Bee	80	2000	30,000	N/A[1]	N/A	N/A
Century	60	3800	6,500	3.0	2.0	70 (G)
Deep Root	80	3800	155,000	N/A	7.1	68 (R)
Shawtown	80	7400	13,200	400,000	8.5	102 (R)
Vespro[2]	70	2000	30,000	N/A	N/A	N/A

[1] Datum not available.

[2] Currently retooling to produce a different style barrier.

information is based on the results of plastics engineering tests run in accordance with American Society for Testing and Materials (ASTM) guidelines (1). But what do these test data reveal to consumers? The ASTM guidelines repeatedly state that results of stress and flexural tests conducted on plastics under laboratory test conditions do not indicate that the same relationships will exist under temperatures and other environmental parameters widely different from those of the test conditions. This is because of the high degree of sensitivity of many plastics to rate of straining under different environmental conditions. Nor are impact test results generally considered a measure of the abrasion or wear resistance of these plastic materials.

Overall, the significance and use of these tests is for quality control and specifications purposes during production (1). Data derived from these tests and reported by barrier manufacturers in their advertising brochures and product labels do not provide consumers with information on how the barriers will resist the wear and tear of daily exposure to foot-traffic, landscaping equipment or other performance features. Until better information is available, on-site observations of barrier performance, as reported in Table 3, will be critical in sorting out which barriers best meet particular purposes.

A barrier's effectiveness is nullified if the top edge is not permanently visible. Roots have readily overgrown barriers in experiments at SUFRA that were accidentally covered with even thin layers of soil (Barker, unpublished data). We see this same problem in commercially installed landscaping anytime soil or organic mulch of any depth obscures the top edges of root barriers (Fig. 2). Similarly, it is vital that barrier panels not pull apart or crack because of faulty connectors. It may be no coincidence that the two barriers made of either high-density polyethylene or polypropylene and exhibiting superior sturdiness and durability were injection molded. Predisposition of all of the barriers to above-grade damage by foot-traffic and landscaping equipment may be markedly altered by designing a wider, more durable top edge on the barriers, a feature that would require that they be made by injection molding instead of by extrusion. Manufacturers are aware of the above-grade wear and tear problem, and some are aggressively addressing it. Deep Root has recently started marketing a new barrier with a sturdy 7/16-inch wide top edge designed to support foot traffic and retard root overgrowth. Controlled longevity of

Figure 2 — Tree roots overgrowing a root barrier where the top edge had been covered by a thin layer of organic mulch.

Table 3. Comparison of five root barriers based on observations during and after their installation, listed alphabetically by manufacturer

Barrier manufacturer	Installation instruction	Maintains shape while being installed	Panel slippage or settling during or after installation	Connector slippage, settling, or failure	Possible problem[1]
Bumble Bee	In sales brochure only	Yes	No	No	
Century	In sales brochure only, no instructions for use of bonding agent	Yes	No	No	Health risk of inhaling chloromethane bonding agent, comparatively short life expectancy of polystyrene material exposed to sunlight
Deep Root	Printed on shipping carton and on inside of each panel	Yes	Yes	Must be adjusted during installation	
Shawtown	Printed on shipping carton and on inside of each panel	Yes	No	No	Comparatively short life expectancy of polystyrene material exposed to sunlight
Vespro[2]	In sales brochure only	No	Yes	Pulled apart, once installed	Highly malleable, easily loses circular shape; connector failure and above-ground portion of panels occasionally rip vertically

[1] Besides possible breakage of above-ground portion of barrier by foot traffic, mower wheels, or other impacts.
[2] Currently retooling to produce a different style barrier.

root barriers is still another attribute needing attention (4).

In summary, though there still is no clear evidence that the use of root barriers on street trees does not harm the trees and does, indeed, reduce sidewalk damage, root barriers currently on the market, including those herein described, are providing urban tree managers, landscape architects, and home owners with options for dealing with tree roots that grow differently than desired. Moreover, root barriers still are being improved, driven by experiences of consumers and manufacturers and by rigorous experimentation.

Acknowledgment

Use of trade or firm names in this paper is for reader information and does not imply endorsement by the U.S. Department of Agriculture of any product or service.

Literature Cited

1. American Society for Testing and Materials. 1992. Annual Book of ASTM Standards, Vol. 08.01, Designations D 256-90b (pp. 58-74.), D 638-91 (pp. 159-171), D 785-89 (pp. 252-256), and D 790-91 (pp. 269-278). Philadelphia, PA. American Society for Testing and Materials.
2. Barker, Philip A. 1983. Some urban trees of California. Maintenance problems and genetic improvement possibilities. In Gerhold, Henry D., ed. METRIA 4, Proceedings of the fourth biennial conference of the Metropolitan Tree Improvement Alliance, 1983 June 20-21, Bronx, NY. University Park, PA. The Pennsylvania State University, School of Forest Resources; 47-54.
3. Benavides Meza, Hector M. 1992. Current situation of the urban forest in Mexico City. J. Arboric. 18(1):33-36.
4. Clendinning, Robert A., Steven Cohen, and James E. Potts. 1974. Biodegradable containers: Degradation rates and fabrication techniques. In Tinus, Richard W., William I. Stein, and William E. Balmer (eds.), Proceedings of the North American Containerized Forest Tree Seedling Symposium, Denver, Colorado, August 26-29, 1974, Great Plains Agricultural Council Publication No. 68, pp. 244-254.

Interactions Between Aeration and Moisture Content in Selected Urban Soils

Laurence R. Costello, James D. MacDonald and Terrence Berger

It is recognized that root zone aeration may be impaired by a number of soil factors, including high moisture levels, compaction, grade changes, surface barriers (concrete, asphalt, etc.) or combinations of the above (1). Among these, high soil moisture is probably the most ubiquitous and frequent cause of aeration deficits (2,4). Soil moisture levels may be excessive for short or long periods of time directly as a result of frequent and/or heavy irrigations or prolonged rainfall, and indirectly from high soil bulk density, poor soil structure (a high ratio of micropore space to macropore space), fine soil texture, subsurface compacted layers (hardpans or plowpans), and shallow water tables. Certain tree species are tolerant of high soil moisture (e.g., *Taxodium distichum, Salix* spp., and *Quercus bicolor*), while others are not (*Fagus sylvatica, Prunus avium,* and many *Quercus* spp.). Most species lay between the extremes and, because soil moisture is often a controllable component of urban soils (through irrigation management), tree performance may depend upon proper irrigation (i.e., aeration) management.

Although information has been published on the impacts of high soil moisture on aeration in agricultural soils (e.g., 2,4), little data exist for urban soils. Our objective in this study was to examine the relationship between soil moisture content and aeration in selected landscape soils.

Methods and Materials

Six field study sites were selected on the Davis campus of the University of California. Four were landscaped sites (irrigated turf) with automatic irrigation systems and two were fallow sites that had provision for manual irrigation. Site descriptions and soil characteristics are presented in Table 1.

Each site was monitored at various times (for intervals ranging from 7-17 days) during the growing season for soil moisture and aeration status. Soil aeration status was measured using the oxygen diffusion rate (ODR) method (5). Soil moisture was monitored using tensiometers (Soil Moisture Equipment Corp, Santa Barbara, CA). The equipment and procedures for obtaining ODR measurements, soil moisture measure-

Laurence R. Costello is with the University of California Cooperative Extension, 625 Miramontes, Room 200, Half Moon Bay, CA 94019. James D. MacDonald and Terrence Berger are with the Plant Pathology Department, University of California, Davis, CA 95616.

Table 1. Site description and physical properties of field study soils

Site	Description	Textural class	Bulk density
1	Lightly-trafficked turf parking strip with chlorotic cork oak (20+ yrs. old)	Clay loam	1.56
2	Lightly-trafficked turf parking strip with vigorous cork oak trees (20+ yrs. old)	Sandy clay loam	1.40
3	Heavily-trafficked turf area around declining coast live oak (50-75 yrs. old)	Sandy clay loam	1.57
4	Lightly-trafficked turf parking strip with vigorous cork oak (20+ yrs. old)	Clay loam	1.50
5	Fallow field area with irrigation controlled manually	Clay loam	1.50
6	Fallow field area with irrigation controlled manually	Clay	1.38

ments, and soil characterization analyses have been described previously (6).

Although measurements were obtained at three different soil depths (10-15 cm, 25-30 cm, and 90-100 cm) at each site, only data collected from the 10-15 cm depth are reported here. This is believed to be the depth at which most root development and activity occurs (3) and where changes in soil moisture and aeration status are likely to be greatest.

Results and Discussion

Soil moisture content and oxygen diffusion rate were found to be closely related at all sites. Generally, as soils dried ODR increased, and as soils became wet ODR decreased (e.g., Figs. 1, 2). In some cases where moisture content stayed high for extended periods, ODR never reached favorable levels (e.g., Fig. 3). The ODR level below which root function of a plant is severely impaired can be defined as the critical ODR, or ODRc. Oxygen diffusion rates ≤ 0.2 μg/cm^2/min have been found to impair root function in many plant species (8). Although there is probably wide variation in ORCc among species, 0.2 μg/cm^2/min is considered a good approximation of the critical ODR level for most plants.

At site 1 (Fig. 1), the soil was initially wet (<10 cb) and ODR was correspondingly low (<0.3 μg/cm^2/min). After day 4, a drying phase occurred (when the irrigation system was turned off) until the ninth day when matric tensions reached 55 centibars (cb). ODR values increased from 0.2 to 1.0 μg/cm^2/min in a manner that paralleled the increase in matric tension. After day 9 an irrigation episode occurred and both matric tension and ODR declined abruptly. The relationship between moisture content and ODR in this soil suggests that periodic drying cycles are needed for ODR values to increase appreciably

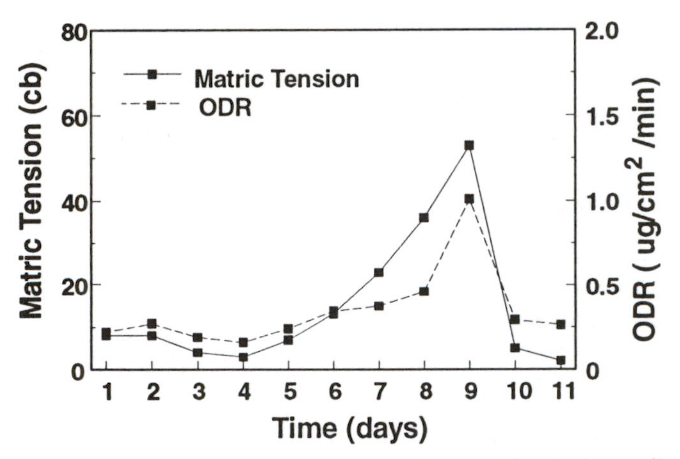

Figure 1 — Moisture content and ODR measurements for clay loam soil at site 1. Increase in matric tension and ODR after day 4 resulted from shut down of irrigation system. Irrigation turned on again after day 9.

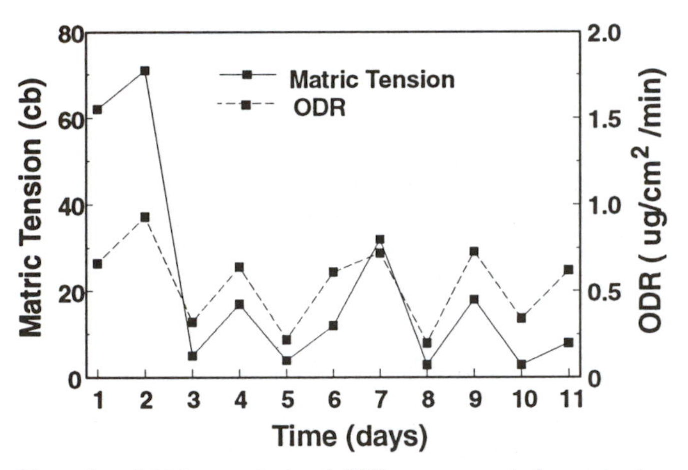

Figure 2 — Moisture content and ODR measurements for sandy clay loam soil at site 2. Irrigations occurred after readings on days 2, 4, 7, and 9.

above the critical ODR level of 0.2 μg/cm²/min. When the soil is maintained in a wet condition (<10 cb) for prolonged periods, ODR values remain very close to the ODRc.

Greater cycling of soil moisture and ODR was found at site 2 (Fig. 2). Here, multiple wetting and drying episodes occurred over the 11 day monitoring period. Initially, the soil was relatively dry (day 1 and 2). Then an irrigation occurred (day 3) and matric tension dropped from 70 cb to 5 cb and ODR declined from 1.0 to 0.35 μg/cm²/min. Subsequent irrigations on days 5, 8, and 10 showed a similar relationship between matric tension and ODR. Unlike site 1, soil moisture fluctuated frequently at this site and, because ODR also fluctuated, there was a substantial accumulation of time where ODR levels were above the ODRc.

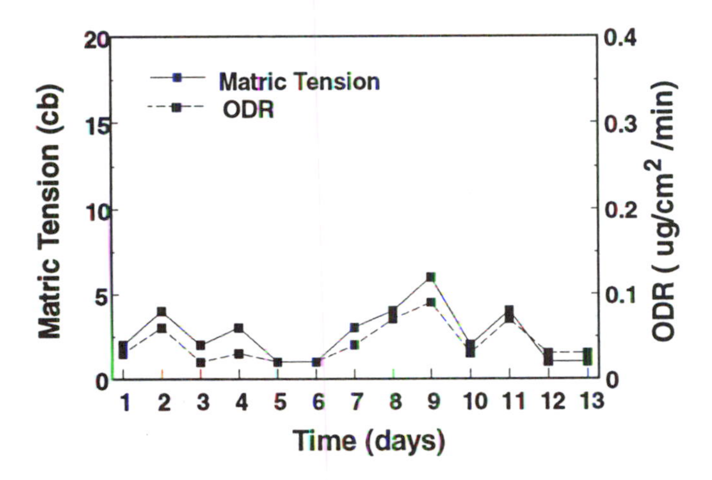

Figure 3 — Moisture content and ODR measurements for sandy clay loam soil at site 3. Note expanded scale for matric tension and ODR relative to other figures.

At site 3, moisture levels were very high (<5 cb) for the entire measurement period, and ODR values were commensurately low (≤ 0.1 μg/cm²/min) (Fig. 3). Since there were no significant drying episodes, ODR values stayed well below the ODRc. However, in addition to high soil moisture, the soil at this site also had a bulk density sufficiently high (Table 1) to be considered restrictive (7). Thus, ODR levels were probably limited by both the moisture content and bulk density, and a drying cycle alone may not have been sufficient to increase ODR levels in this soil. It was clear, however, that the oak at this site was declining and there was very little cumulative time where ODR was above the ODRc.

At site 4, on the other hand, the soil was very wet (<10 cb) throughout the monitoring period (10 days), but ODR levels were consistently above the ODRc (Fig.4). Presumably at this site, soil structural properties were adequate to allow soil drainage and gaseous diffusion. The tree at this location was vigorous and drying cycles apparently were not necessary to maintain ODR levels above the ODRc.

A similar pattern, where moisture levels were high and yet ODR levels were greater than the ODRc, was found at site 5 (Fig. 5). Here, matric tensions were maintained at approximately 10 cb for a 17 day period. While ODR values declined from 0.6 to 0.3 μg/cm2/min, they did not drop below the ODRc. The moisture release characteristic of this soil showed that there was adequate drainage and open pore space at moderate tensions allowing for gaseous diffusion.

When moisture content was maintained at a high level (<10 cb) in clay soil (site 6), ODR was very low (0.1 μg/cm²/min). Little drying occurred in this soil, even when irrigation inputs ceased. Therefore, unlike the soil at sites 4 and 5, drainage at 10 cb matric tension was not sufficient to allow the ODR level to rise above the ODRc for any time during the 7 day study period.

Collectively these results indicate that oxygen diffusion rates in soils are strongly affected by moisture content. Generally, as a soils dries the ODR increases, while a wetting phase causes the ODR to decrease. The magnitude and duration of the effect was found to vary, however, with soil texture and bulk density. Soils of relatively coarse tex-

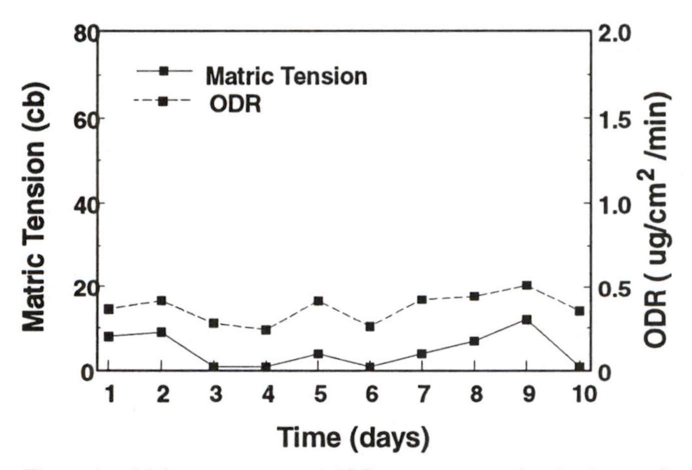

Figure 4 — Moisture content and ODR measurements for clay loam soil at site 4. ODR measurements remained relatively high even though matric tension was low.

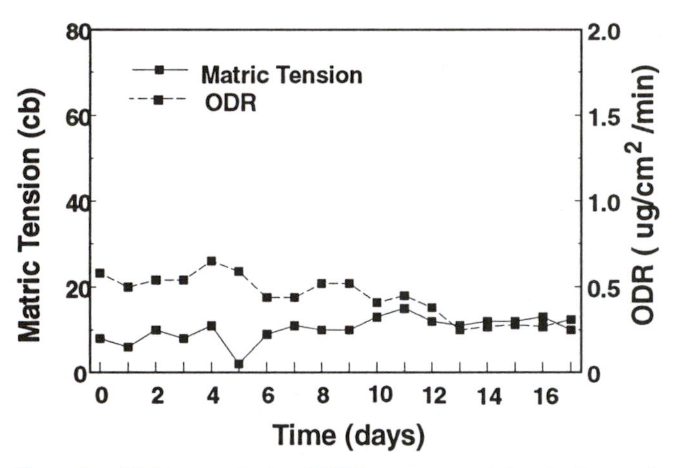

Figure 5 — Moisture content and ODR measurements for clay loam soil at site 5. This was a fallow soil where irrigation was controlled manually.

ture (and presumably good structure) sustained high moisture levels without reducing ODR below the ODRc for extended periods (eg. sandy clay loam at site 2 or silt loam at site 5). In soils of finer texture, however, ODR levels were depressed below the ODRc for the duration of the high moisture episode (eg. clay at site 6 and clay loam at site 1). Restrictive bulk densities at sites 1 and 3 likely contributed to low ODR levels. Cycling of water and air was found where the drying phase between irrigations was of sufficient duration to allow moisture content to decrease and ODR levels to increase (eg.site 2).

From a management perspective, these results suggest that favorable aeration conditions likely can be sustained in many soils by allowing a drying phase after a wetting

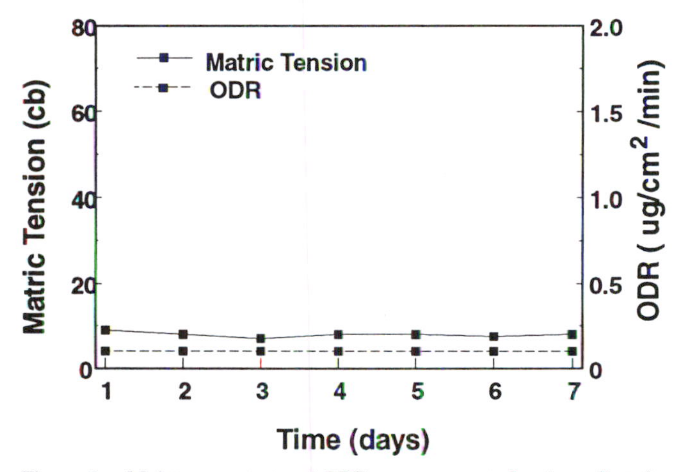

Figure 6 — Moisture content and ODR measurements for clay soil at site 6. Soil remained wet and ODR low for the 7 day study period. Irrigation was controlled manually.

episode. Soils of finer texture and/or poor structure will need a longer drying phase than soils with coarse texture and good structure. In some very coarse soils (eg. sands) favorable aeration levels may be easily achieved, while maintaining adequate moisture levels will be a greater challenge.

These results emphasize the importance of understanding soil physical properties and their effects on air and water levels in soils. Maintaining favorable balances between soil aeration and moisture content requires some knowledge of soil texture, structure, bulk density, and depth. This information will assist landscape managers in determining for instance how long a drying phase is needed to maintain adequate aeration levels or how frequently to irrigate to maintain adequate moisture levels. As a result, soil conditions favorable for root function, growth, and health can be maintained for optimal plant performance.

Literature Cited

1. Craul, P.J. 1992. Urban Soil in Landscape Design. John Wiley and Sons, New York, NY.
2. Glinski, J. and Stepinewski, W. 1985. Soil Aeration and its Role for Plants. CRC Press, Boca Raton, FL.
3. Himelick, E.B. 1986. Root development of trees growing in an urban environment. Illinois Natural History Survey Rep. No. 262, December.
4. Kozlowski, T. T. (ed.). 1984. Flooding and Plant Growth. Academic Press, Orlando, FL.
5. Letey, J. and Stolzy, L.H. 1964. Measurement of oxygen diffusion rates with the platinum microelectrode. I. Theory and equipment. Hilgardia 35:545-554
6. MacDonald, J.D., Costello, L.R. and Berger, T. 1993. An evaluation of soil aeration status around healthy and declining oaks in an urban environment in California. J.Arboric. 19 (4):209-219.
7. Morris, L.A. and Lowery, R.F. 1988. Influence of site preparation on soil conditions affecting stand establishment and tree growth. South J. Applied For. 12:170-178.

8. Stolzy, L. H. and Letey, J. 1964. Measurement of oxygen diffusion rates with the platinum microelectrode. III. Correlation of plant response to soil oxygen diffusion rates. Hilgardia 35:567-576.

Using Steel Rods for the Non-destructive Assessment of Aeration in Urban Soils

Simon J. Hodge

Diagnosis of poor performance in urban trees has to be based on a knowledge of soil physical conditions. Soil aeration status can be inferred from the corrosion of steel rods inserted into the ground. The technique is inexpensive, non-destructive and can provide information on the vertical distribution of anaerobic conditions for visual or statistical analysis. A case study is presented which shows how the technique was used to assess the effectiveness of compressed air soil injection.

Non-structural roots are composed of living cells which need oxygen to survive and, for most tree species, this oxygen must come from the soil. The diagnosis of poor amenity tree performance often requires the investigation of soil aeration beneath tarmac or paving, and this presents serious technical problems. Conventional methods using an oxygen probe or platinum micro-cathodes each have shortcomings particularly in the urban environment which is characterized by compacted stony and variable soils, hard pavings, and the risk of human interference. Forestry Commission research has developed the steel rod technique to assess soil aeration in these situations.

The Steel Rod Technique

In most circumstances, oxygen status is the soil factor that influences most the corrosion of mild steel (8). Using this fact, Forestry Commission researchers used steel rods for the prediction of rooting depth on upland soils (1). By recording the depth of rusting on mild steel rods, the depth of aerated soil, and hence the likely depth of tree rooting could be predicted. The UK Department of the Environment began funding research using steel rods to evaluate urban soils in 1987 (2). Tools were developed for rod insertion and extraction in the urban environment and a more detailed form of assessment and analysis was devised to accommodate the vertical variability of urban soils.

A detailed description of the procedure for using steel rods is found in Hodge & Knott (7) and a code of safe working is detailed in Appendix 1. Rods are 60 cm long, 6 mm in diameter and made of bright mild steel (non-leaded, low sulphur content). One end is pointed and the other has a grove machined to 1 mm depth to aid extraction. Rods are inserted for three months. The amount of information obtained is enhanced if two sets of rods are inserted, corresponding to spring and summer. The number of rods inserted into the rooting zone of each tree depends on tree size, site

Simon J. Hodge is with the Forestry Authority Research Division, Alice Holt Lodge, Farnham, Surrey, GU10 4LH, England.

variability and the number of trees in the group being studied. Investigating the soil around a single, large urban tree may require 10 rods, whilst investigating poor performance of a newly planted avenue on a homogenous soil may justify only two rods per tree. Rods must be wiped clean of any engineering oil directly before insertion, then driven vertically into the ground. A tool has been developed to do this. Rods can be driven through thin layers of tarmac or hardcore (crushed stone) but to insert rods through more substantial surfaces, holes must be pre-drilled with a 6.5 mm bit to the depth of the surface material. The location of each rod must be mapped as public access to most sites means that rods cannot be flagged. Two tools have been developed to extract rods vertically to ensure that the patterns of corrosion are not damaged.

After extraction the rod is immediately swabbed with a fabric cloth soaked in a 10% (v/v) solution of ammonia to remove soil and to stop further rusting. Secondary rusting starts very quickly after extraction so only a few rods are extracted at a time before swabbing and, if possible, rods should not be extracted during wet weather. Once swabbed, each rod is labelled and kept in a dry atmosphere.

If rods are being inserted for a subsequent season, their location must be 10 cm away from the hole of the previous rod to ensure good contact with previously undisturbed soil.

Interpretation of Corrosion Patterns

Carnell and Anderson (1) identified five types of surfaces that may be encountered on steel rods extracted from the soil. Interpretation of corrosion types, which was clarified in a laboratory study (6) is:

***Red/brown Rust*:** Indicates a well aerated soil.

***Raised Black*:** Occurs where rusting has started but has been interrupted, or where rust has been knocked off during removal of the rod from the ground.

***Shiny Metal*:** If occurring in association with soils containing substantial undecomposed organic matter, it is indicative of conditions hostile to rooting. Corrosion is prevented by the presence of organic compounds such as polyphenols produced from organic matter under anaerobic condition. In mineral soils, particularly drought prone coarse textured soils, the presence of substantial amounts of shiny metal cannot be used as an indicator of suitability of root growth.

***Smooth Black*:** Occurs where anaerobic bacteria utilize soil sulphates producing hydrogen sulphide, which reacts with the surface of the metal.

***Matt Grey*:** Indicates totally anaerobic conditions.

Assessment and Presentation of Corrosion Patterns

For Visual Analysis

Rods from relatively undisturbed soils can be assessed by determination of the maximum depth of substantial red/brown rust and raised black (indicating condition conducive to root growth) or the depth of onset of substantial matt grey and smooth black (indicating conditions that are hostile to root growth). However, on disturbed urban sites such zones may not be readily discernable and a more detailed form of assessment is required.

The rod is laid along a meter rule and an assessment made of the two categories; matt grey and smooth black combined, and shiny metal. The presence of the two categories in patches of more than 0.5 cm along a line down the rod is recorded for each

3 cm section of rod. When corrosion down the first line has been recorded, the rod is turned through 180° and the process repeated.

The data must then be translated into numerical form. The information is recorded as a score of 0 to 12 (6 x 0.5 cm on each side of the rod) for each 3 cm section. Initially scores for matt grey/smooth black and shiny metal can be combined to form a score for inhospitable soil conditions. However, if results prove inconclusive, and the presence of shiny metal is substantial, the scores for matt grey/smooth black alone can be used for a reconsideration of the profiles.

The comparison of the corrosion profiles between seasons has proved particularly useful and Fig. 1 shows the mean rod assessed at 3 cm intervals over three seasons for a paved and gravel area. The profiles from the paved area show seasonal waterlogging, while those from the gravel area indicate the presence of a compacted layer between 15 and 30 cm.

For Statistical Analysis

Evaluation and analysis of steel rod corrosion patterns from a field study (5) showed that data can be collected in four 15 cm sections down the rod without significantly reducing the sensitivity of statical analysis compared to data collected in 3 cm sections. For statistical analysis the total anaerobism score for each rod and the total anaerobism score from 0 to 30 cm minus total score from 31 to 60 cm for each rod were found to be reliable values in expressing the aeration status of the soil (6).

These parameters can be used to compare soil aeration status by analysis of variance over space: between treated or untreated areas; between ailing trees and healthy trees within a group; and over time; 1) periodically after site development or reinstatement; and 2) before and after soil treatments.

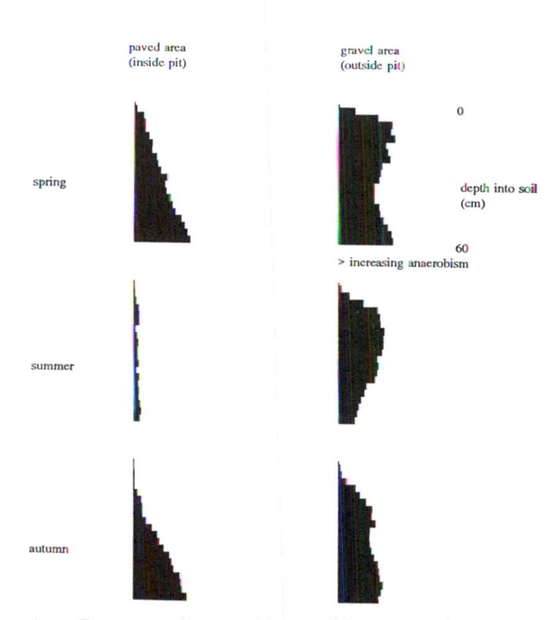

Figure 1 — Presence of anaerobic conditions over three seasons in a paved area and a compacted gravel area.

A Case Study

Compressed air injection is being advocated as a means of relieving compaction in the soil around established trees. Four experiments were set up by the Forestry Authority Research Division to assess the effectiveness of this technique (3,4). The steel rod technique was used to assess the effect of compressed air injection on the soil in order to complement the assessments of tree response.

In the two experiments set up in 1987, compressed air was injected into clay soil around 12 to 16 year old horse chestnut (*Aesculus hippocastanum*) and London plane (*Platanus acerifolia*). No significant treatment effect was detected after injection around the horsechestnut, but London plane did show a 35% increase in shoot extension after two growing seasons. The steel rod technique was used to gain an impression of soil conditions, but at this time effective means for statistically analysing the results had not yet been developed. The steel rods, inserted for three months over the summer, indicated that compressed air injection had relieved a compacted layer in the soil between 10 and 30 cm (Fig. 2).

In an experiment set up in 1991, an avenue of 56 sweet chestnut (*Castanea sativa*), (age 50, average dbh 55 cm, average height 10.3 m) which were slow growing and dying back, were selected. They were located on a wide grass verge on either side of a busy road where the clay loam soil had been moved, reconstituted and compacted. Compressed air injection was undertaken by a professional operator. No statistically significant positive effects on tree condition and growth were detected after two growing seasons.

The steel rod technique was used to assess soil aeration around treated and untreated trees. Four rods per tree were inserted 1.0 m apart and 1.5 m from the tree stem for three months in the winter of 1991-92 and for three months in the summer of 1992. Patterns of corrosion were assessed in 15 cm sections to determine soil aeration status to 60 cm. The total anaerobism score for each rod and the total anaerobism score from 0 to 30 cm minus the total score for 31 to 60 cm were calculated and analysis of variance undertaken on each of these parameters. There were no significant differences between treatments. The corrosion scores were then subjected to principal component analysis to see whether the variation in corrosion scores between rods could be better explained by other patterns of rod corrosion. No patterns could be discerned in relation to compressed air injection treatments for either the summer or winter rods.

The results were validated by dry bulk density assessment with a gamma probe. Assessments were made at 20 cm depth, 1.0 m from the stem of each tree. Sixteen values were recorded, eight by trees not receiving the compressed air treatment and eight by injected trees. The mean dry bulk density of both the treated and untreated ground was 1.7 g/cm^3 and an independent sample t-test confirmed that there was no significant difference in values between the treatments.

Conclusion

The steel rod technique is a practical and relatively inexpensive way of assessing the aeration status of urban soils. The technique can be used to examine the distribution of anaerobic conditions down through a soil. Comparison of rod information between seasons can show whether anaerobic conditions are due to waterlogging or

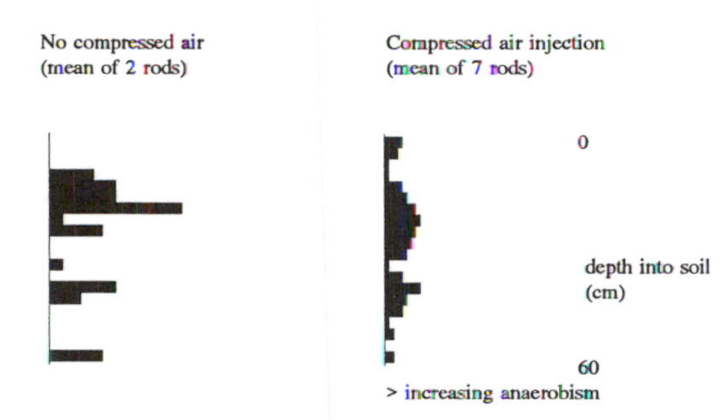

Figure 2 — Distribution of compaction in a compressed air injected soil (From 3).

compaction. The technique can be used to assess aeration under hard surfaces without disturbing the site. The ability to depict visually the condition of soil that cannot otherwise be observed is a compelling advantage of this technique.

Detection of the cause of decline of mature urban trees, or slow growth of young trees, is notoriously difficult. The steel rod technique can provide information on soil conditions that allows precise targeting of remedial treatments such as compressed air injection.

Literature Cited

1. Carnell R. and M.A. Anderson 1986. A technique for extensive field measurement of soil anaerobism by rusting of steel rods. Forestry 59(2): 129-140
2. Colderick S.M. and S.J. Hodge 1991. A study of urban trees. **In** Research for Practical Arboriculture. Forestry Commission Bulletin 97 ed. S.J. Hodge, 29-37. HMSO, London.
3. Hodge, S.J. 1991. Improving the growth of established amenity trees: site physical condition. Arboriculture Research Note 102/91/ARB. Arboricultural Advisory and Information Service, Farnham.
4. Hodge S.J. 1993. The effect of stem nutrient injection and compressed air soil injection on the performance of established amenity trees. Arboric. J. 17(3): 287-300.
5. Hodge, S.J. and R. Boswell 1993. The relationship between site conditions and urban tree growth: an intensive study. J. Arboric. 19(6): 358-367.
6. Hodge, S.J., R. Boswell, and K. Knott. 1993. Development of the steel rod technique for the assessment of aeration in urban soils. J. Arboric. 19(5): 281-288.
7. Hodge, S.J. and K. Knott 1993. A practical guide to the use of steel rods for the assessment of aeration in urban soils. J. Arboric. 19(5): 289-294.
8. Hudson, J.C. and K.C. Watkins. 1968. Tests on the corrosion of buried cast iron and mild steel pipes. BISRA open report MG/B/3/68, C/6/68. London.

Appendix 1: A code of safe working for the use of steel rods in urban areas.

Before starting to insert rods:

i. Obtain a copy of relevant safety manuals for working around underground services (in the UK: Health and Safety Executive Guidance Note 47 "Avoiding danger from underground services").
ii. Obtain maps of underground services (electricity, telephone, gas, cable television, water and sewerage) in the area of work.
iii. Use a cable avoidance tool to locate and check the position of electricity cables on the site.
iv. Check the position of other services on the site.
v. Have contact telephone numbers for underground service companies at hand.
vi. *If there are any doubts as to the precise location of underground services, do not proceed.*

When inserting rods:

vii. Work in pairs.
viii. Keep the public away from the area of work.
ix. Wear rubber soled boots and high voltage specification rubber gloves (in the UK: 4000 volts (working), conforming to British Standard 697:1986).
x. Insert rods using insertion tool.
xi. Ensure that the ends of inserted rods, if left protruding, are not a hazard to the public.

When extracting rods:

xii. When using ammonia solution to swab extracted rods wear protective goggles, boots, rubber gloves and waterproof coverall.
xiii. Splashes of ammonia solution on the skin and clothing should be swabbed with clean water.
xiv. Carry an eyebath and bottle of eye wash in case of ammonia solution splashing into the eyes.

The Effects of Soil Aeration Equipment on Tree Growth

E. Thomas Smiley

Urban soils are prone to soil compaction primarily due to pedestrian and vehicular traffic, and construction activities. Compaction can be a significant limiting factor for plant growth due to the adverse effects on root development. Effective treatments to alleviate soil compaction on existing trees must rapidly modify soil bulk density without causing significant root disturbance. The Grow Gun and Terralift are two machines developed in recent years to meet these criteria. Previous studies of these machines have focused on operation and effects on the soil (2). The second and final phase of our study of aeration equipment focuses on the long term growth response of trees treated with soil aeration machines and conventional soil compaction treatments.

Materials and Methods

Four species of trees were selected at three sites for treatment. All trees were planted two or more years prior to treatment and were growing poorly. All sites had clay loam or clay soils and bulk densities greater than 1.5 which are considered highly compacted (Table 1).

Treatments were applied in February and March 1989. They included the Terralift and Grow Gun soil aeration machines, lawn type core aerator and soil injection of water through a fertilizer injection needle (Table 2). Various dry materials were injected

Table 1. Description of research sites at time of treatment

Name	Tree species	Mean soil moisture (%)	Mean bulk density (g/cc)
Charlotte Douglas Airport Charlotte, NC	Willow oak (*Quercus phellos*)	18	1.52
Cherry Park Rock Hill, SC	Sweetgum (*Liquidambar styraciflua*)	24	1.58
	Red maple (*Acer rubrum*)		1.52
Airport entrance freeway Charlotte, NC	Aristrocrat (*Pyrus calleryana* cv. 'Aristrocrat')	15	1.67

E. Thomas Smiley is with the Bartlett Tree Research Laboratory, 13768 Hamilton Road, Charlotte, NC 28278.

through the soil aeration machines to maintain the openings created by the compressed air discharge. These included Styrofoam 'C' beads, Styrofoam insulation beads, Supersorb 'F' and perlite. When more than one material was injected with a single machine, and the results were not significantly different, the results were combined.

Bulk density was determined by collecting two-inch long by two-inch diameter cores of soil, oven drying them at 105°C, and weighing them.

Diameter of trunk growth was measured annually using a diameter tape at a marked point on each tree. Results were statistically analyzed using an analysis of variance.

Results and Discussion

No treatments demonstrated a consistent or statistically significant increase in diameter growth compared with the control trees (Fig. 1-4). This is due to the small amount of soil that was affected by the soil aeration treatment. Previous work showed that only a single fracture was not enough to reduce bulk density in the rooting zone of the tree.

These results are consistent with Rolf (1) where he found a reduction in bulk density in the sandy loam site but not in the loam site. However, typically there is more concern with compaction in loam and clay loam sites than on sandier sites.

Soil aeration machines may also produce acceptable results when a perched water

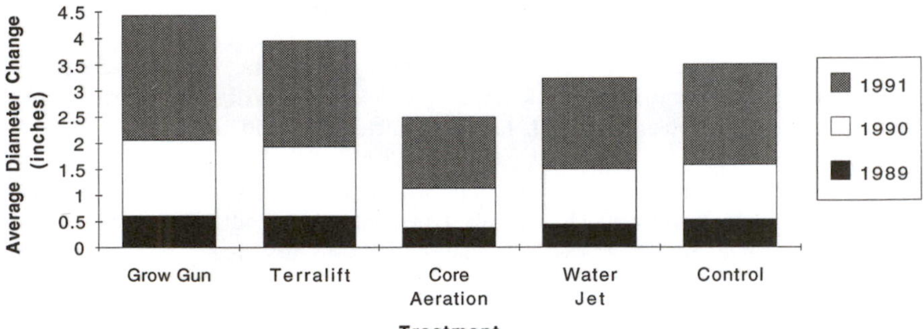

Figure 1 — Diameter change in willow oaks (*Quercus phellos*) over three years after treatment with various machines to reduce soil compaction, no values differ significantly.

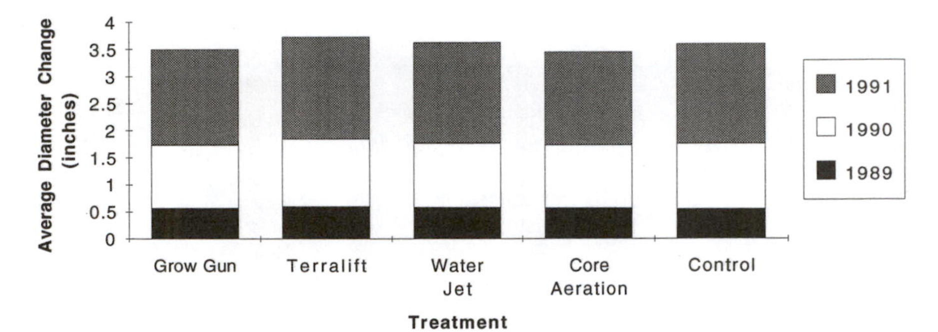

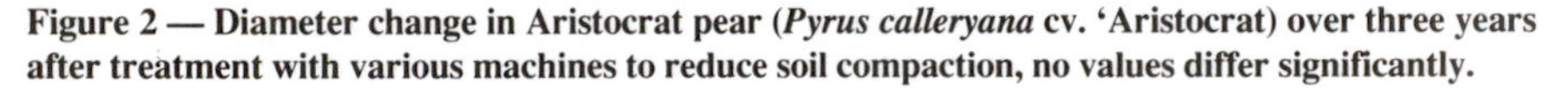

Figure 2 — Diameter change in Aristocrat pear (*Pyrus calleryana* cv. 'Aristocrat) over three years after treatment with various machines to reduce soil compaction, no values differ significantly.

Table 2. Soil aeration treatments

Treatment code	Number of trees	Machine	Probe depth (cm)	Fill material	Fill vol. (liters/ fracture)	Fractures per tree
Site: Douglas Airport - Willow oak						
G Sty	5	Grow Gun	30	Styrofoam insulation	2	4
TL Comb	5	Terralift	45/70	Styrofoam 'C' Supersorb 'F'	2X2 2X0.025	4
Core Ar	5	Core aerator	8	—	—	—
Water	5	Fert. needle	20	Water		6
Control	5	—	—	—	—	—
Site: Airport entrance freeway - Astrocrat pear						
GG Sty	11	Grow Gun	30	Styrofoam insulation	2	4
GG Sup	11	Grow Gun	30	Supersorb 'F'	0.2	4
TL Sty	11	Terralift	45/70	Styrofoam 'C' beads	2X2	3
TL Sup	11	Terralift	45/70	Supersorb 'F'	2x0.075	3
Core Ar	11	Core aerator	8	—	—	—
Water	11	Fert. needle	20	Water	2	6
Control	11	—	—	—	—	—
Site: Cherry Park - Red maple						
TL Per	10	Terralift	45/70	Perlite	2X2	4
Core Ar	5	Core aerator	8	—	—	—
Water	8	Fert. needle	20	Water	2	6
Control	9	—	—	—	—	—
Site: Cherry Park - Sweetgum						
TL Per	10	Terralift	45/70	Perlite	2X2	4
Core Ar	5	Core aerator	8	—	—	—
Water	8	Fert. needle	20	Water	2	6
Control	9	—	—	—	—	—

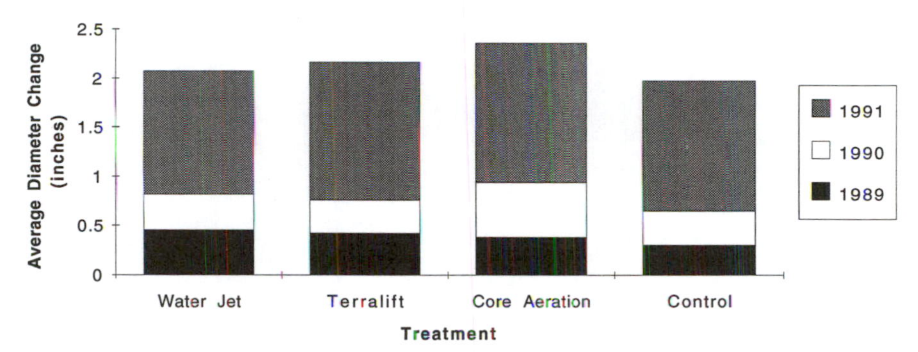

Figure 3 — Diameter change in red maple (*Acer rubrum*) over three years after treatment with various machines to reduce soil compaction, no values differ significantly.

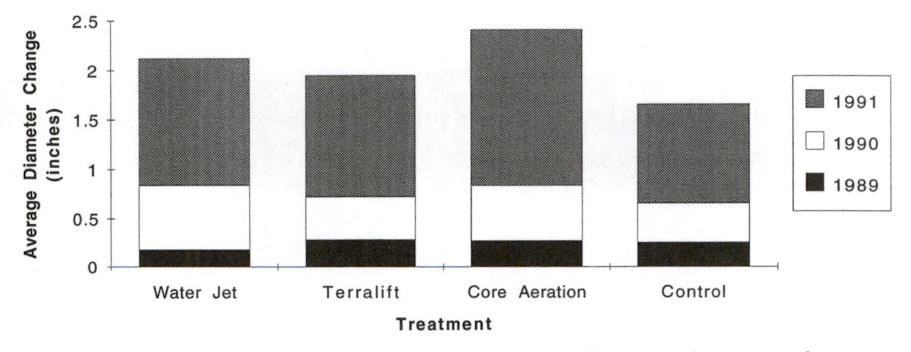

Figure 4 — Diameter change in sweetgum (*Liquidambar styraciflua*) over three years after treatment with various machines to reduce soil compaction, no values differ significantly.

table is created by a hard pan layer, and there is a well drained layer beneath. Drilling through the hard pan then fracturing it with compressed air may improve drainage into the well drained layer beneath. Another treatment which may serve the same purpose is simply drilling a number of holes through the pan layer with a two inch earth auger.

Literature Cited

1. Rolf, K. 1992. Soil physical effects of pneumatic subsoil loosening using a Terralift soil aerator. J. Arboric. 18(5): 235-240.
2. Smiley, E.T., G.W. Watson, B.R. Fraedrich and D.C. Booth. 1990. Evaluation of soil aeration equipment. J. Arboric. 16(5): 118-123.

Evaluating Tree Planter Health: Soil Chemical Perspectives

Patrick Kelsey

Trees in planter boxes endure stressful conditions associated with the physical and chemical soil environment. This study evaluates a number of planter soils in urban settings to assess the impacts of the chemical environment on trees and summarizes soil chemical evaluation techniques and their value in establishing soil conditions in street tree planters. Properties examined include salinity, alkalinity, and oxygen diffusion.

Street trees in planter environments endure harsh physical and chemical soil conditions. Planters inherently restrict rooting volume and their soils often have poor aggregation which reduces root growth (1,5,6). Loss of soil aggregation and subsequent compaction are commonly caused by handling the materials when wet and by using a soil mix with excessively high clay contents. This in turn causes a reduction in available water holding capacity. Lindsey and Bassuk (7) have proposed methods for determining optimum rooting volume to adequately supply trees with water in these situations.

Compaction and water stress are not the only factors limiting growth in urban trees. Alkaline-soil-induced chlorosis from runoff and leaching was the cause of white oak decline adjacent to parking lots with dolomitic limestone bases even though the fills were not initially in the root zones of the trees (10).

Salinity in street tree planters has been shown to restrict plant growth and limit survival of sycamore (*Platanus occidentalis*) and Redmond linden (*Tilia x euchlora* 'Redmond') (4,5). MacDonald et al. (8) found oxygen diffusion rates (ODR) to be a reliable indicator of aeration status over a wide variety of urban sites. Glinski and Stepinewski (2) established reduced rooting viability due to poor aeration when oxygen diffusion rates drop below 0.3 $\mu g/cm^2/min$.

This paper summarizes the findings of a number of case studies of tree planter problems related to soil chemistry.

Study Sites and Conditions

All study sites are in the Chicago, Illinois metropolitan region. The region has naturally alkaline geologic materials. Chicago lies in the heart of the upper midwestern snow belt and has a high frequency of snow and ice events each year that require deicing. Sodium chloride is the deicing chemical most commonly used in the region (IDOT, personal communication).

Patrick Kelsey is a Research Soil Scientist with The Morton Arboretum, Urban Vegetation Laboratory, Route 53, Lisle, IL 60532.

Site 1 — Continental Bank

The Continental Bank site in Chicago consisted of a series of planter holes cut into concrete sidewalk pavement. The planters did not have sides and thus were not technically planter boxes. The soil materials included a mixture of urban rubble: bricks, sand, concrete chunks, cinders, topsoil, and charred wood (remnant of the Great Chicago Fire). The site was excessively well drained. At the time of investigation, Redmond lindens in the planters were in severe decline. The trees had been planted on the site three years prior to the evaluation.

Site 2 — Finkl Steel

The Finkl Steel site in Chicago consisted of very narrow parkways approximately 1 m (3 ft) wide with variable length. The soil materials included steel making debris such as slag, coke, and brick mortar under a thin layer of imported shredded topsoil. The site had also been subjected to highly alkaline solutions used in steel production. The site was excessively well drained. The site supports recently planted hybrid poplars (*Populus* sp.) and elms (*Ulmus* sp.).

Site 3 — City of Geneva

The Geneva sites were sidewalk street tree planters with concrete sides and no bottoms. City utilities were at the base of the planters, some 45 cm (18) inches below the soil surface. All planters in the study area were high in organic matter content and clay and had poor drainage. The planters also contained a large proportion of coarse fragments made up of brick rubble, concrete, and mortar. The sites supported declining sycamores and ginkgo (*Ginkgo biloba*) at the time of study. Following redevelopment of the site and planter soil replacement, Chanticleer pears (*Pyrus calleryana* 'Chanticleer') were planted.

Site 4 — Village of Downers Grove

The Village of Downers Grove maintains planters at grade and raised above grade 45 - 76 cm (18 - 30 inches). Most raised planters had been backfilled with pulverized topsoil at the time of installation. The grade level planters had a large proportion of dolomitic gravel and construction rubble. Planters were generally well drained.

Site 5 — Lake Shore Drive, Chicago, IL

The Lake Shore Drive site was a discontinuous median planter raised 61 cm (24 in) above the pavement. It is 3 m (10 ft) wide and approximately 5 km (3 mi) long. The soils in the planter are highly variable. Most soil samples analyzed were high in organic matter and had clay contents above 27%. The planter was designed with irrigation and drainage. Portions of the planter have a sealed bottom while the remainder lies on top of medium textured beach sand. A wide variety of trees and shrubs have been planted at this site. The entire median planter is centered between eight lanes of high speed traffic.

Methods

Representative soil samples were taken from each planter location. Soils were tested for pH, electrolytic conductivity, cation exchange capacity, base saturation, clay content, sodium, and chloride. Exchangeable sodium and sodium absorption ratio were calculated for each soil. Oxygen diffusion rates were determined in-situ at selected planters in August 1993. The pH was determined electrometrically using a 1:1 soil:water volume ratio (11). Electrolytic conductivity was determined in a saturated paste (12). Cation exchange capacity was determined using 1N NH_4OAc method (11). Elemental analysis was performed on solution extracts using atomic absorption spectrophotometry except chloride which was analyzed by ion specific electrode. Base saturation was calculated as a sum of the bases.

Alkalinity

Alkalinity in soil is controlled by the presence of basic cations (Ca, Mg, K, Na) on the exchange capacity and in the soil solution. Anions causing alkalinity of the soil solution include carbonate (CO_2), bicarbonate (HCO_3^-), and hydroxyl (OH^-) ions. The optimum soil pH range is between 5.5 and 6.2 for most temperate region trees and shrubs. Soils in the Chicago region are slightly acid (5.5 - 6.5) naturally. Conversely, urban soil materials are quite alkaline in most cases (Figure 1). Site 2 shows the greatest variance from the norm with mean a profile pH around 10. This is because the soil is dominated by sodium (Na) and potassium (K) hydroxides. Alkalinity at this level is outside the range in which most plants will grow.

Sites 2 and 4, both grade level planters, show the most common planter alkalinity conditions with pHs ranging from near neutral to pH 8.0 (Figure 1). The dominant source of alkalinity in these instances is the surrounding calcium carbonate-rich infrastructure. Concrete, dolomitic gravel, and mortar all contribute to the alkalinity.

Soil alkalinity in planter pits can not be overcome in urban soil materials because of the large source carbonate-rich materials. Messenger (9) has outlined procedures for chemically acidifying soils in situations where alkalinity is limited, and where there appears to be a limited source of new materials to generate new base cations (Ca, Mg, Na, K), carbonates, and bicarbonates. The most effective way to manage planter soil alkalinity is to replace extremely alkaline materials with slightly acid soil, and then continue a maintenance level of acidification in the planter. This approach requires adequate drainage to leach alkaline compounds as they develop.

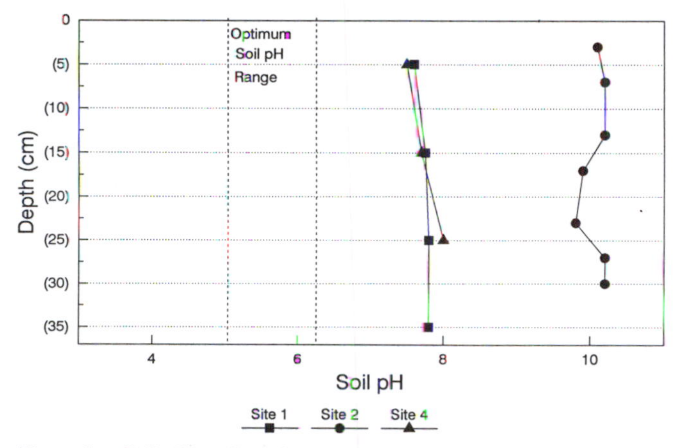

Figure 1 — Soil pH profiles of selected street tree planters.

Salinity

Salinity in the soil can be measured in a number of ways. Electrolytic conductivity (EC), the most commonly used method, measures the relative movement of current through a soil:water paste. This can be correlated directly to the salt content of the soil. Unfortunately, not all salts have the same conductivity. High activity salts like sodium sulfate and sodium nitrate have high conductivities while equal quantities of sodium chloride in a soil paste will give much lower conductivities (3). A number of planter sites with salinity problems were evaluated for conductivity, sodium content, exchangeable sodium percentage, and sodium absorption ratio (Figures 2 and 3). Site 1 had very high conductivities but low exchangeable sodium percentages and sodium adsorption ratios throughout the profile. Thus, the conductivity in the soil was not necessarily related to sodium content but to other high activity salts from fertilization (4). In the Geneva study (Site 3), there was no clear relationship between conductivity and sodium chloride-affected soils (Figure 4). Deicing salt damaged planter soils were better predicted by evaluating the relationship between the exchangeable sodium percentage and the sodium adsorption ratio (Figure 5). A strong and significant correlation ($r = 0.83$) of these was observed in the Geneva planter soils. Base saturation and the associated calculations, exchangeable sodium percentage and sodium adsorption ratio, are more appropriate predictors of salinity problems in urban soils because a low electrolytic conductivity does not rule out the possibility of a salt problem (Figure 2,3 and 6). Low ESP and SAR along with low EC rules out a deicing salt problem (5).

Oxygen Diffusion

Oxygen diffusion rates were low in all profiles examined (Figure 7). There were no clear relationships between oxygen diffusion rates and site soil characteristics. Glinski and Stepinewski (2) determined 0.3 μg/cm²/min to be restrictive to root growth. In general ODRs were above the restrictive value to a depth of only 10 cm and decreased with increasing profile depth. The Geneva pits had the lowest overall ODRs but there were no significant differences between sites. Only the Continental Bank planter pits (Site 1) had

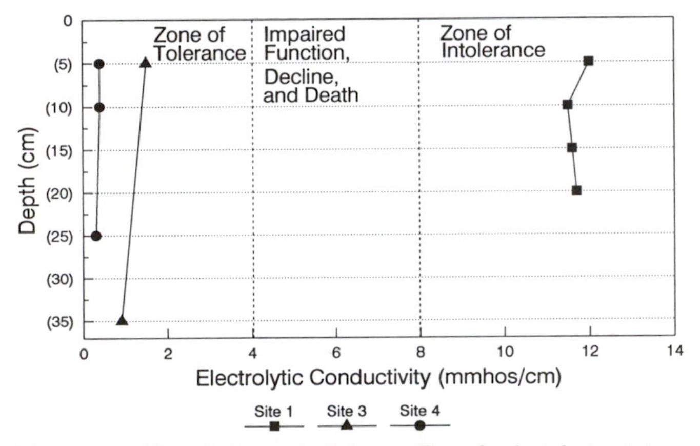

Figure 2— Electrolytic conductivity profiles of selected street tree planters.

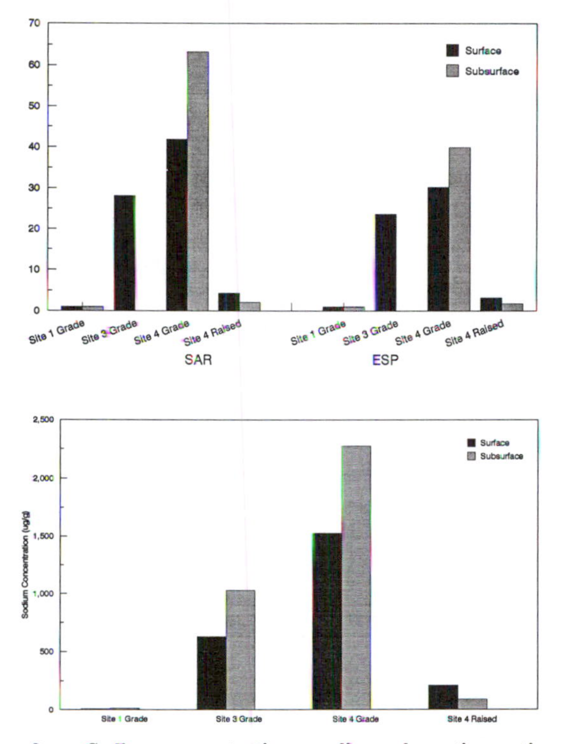

Figure 3 — Sodium concentrations, sodium absorption ratios, and exchangeable sodium percentages for selected street tree planters.

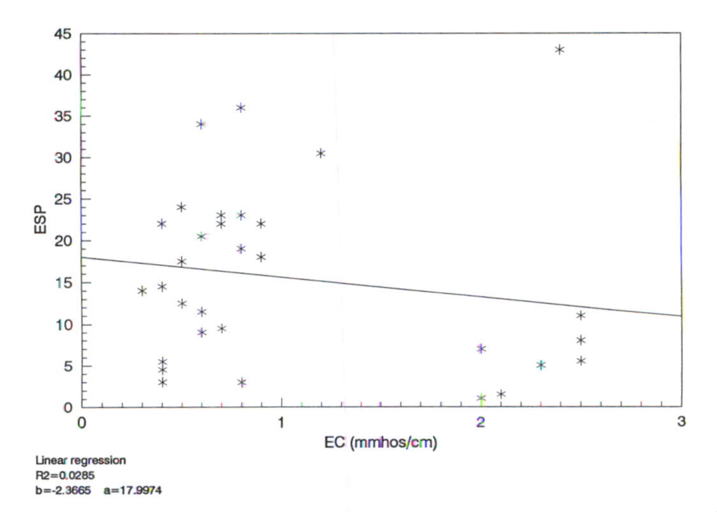

Figure 4 — Relationship between electrolytic conductivity and exchangeable sodium percentage.

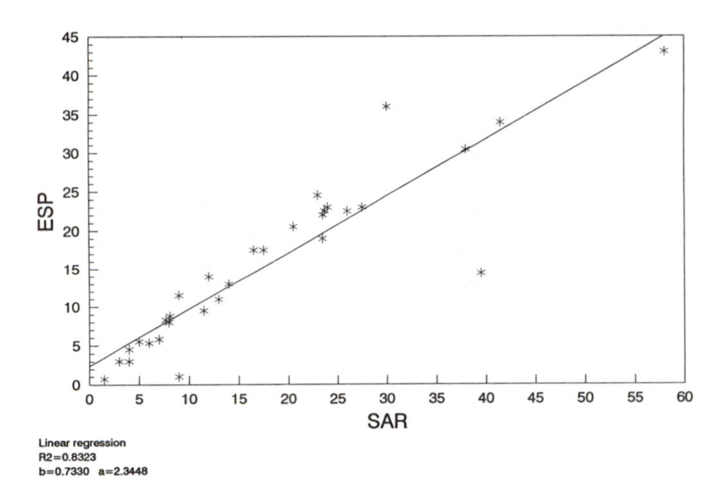

Figure 5 — Relationship between sodium adsorption ratio and exchangeable sodium percentage.

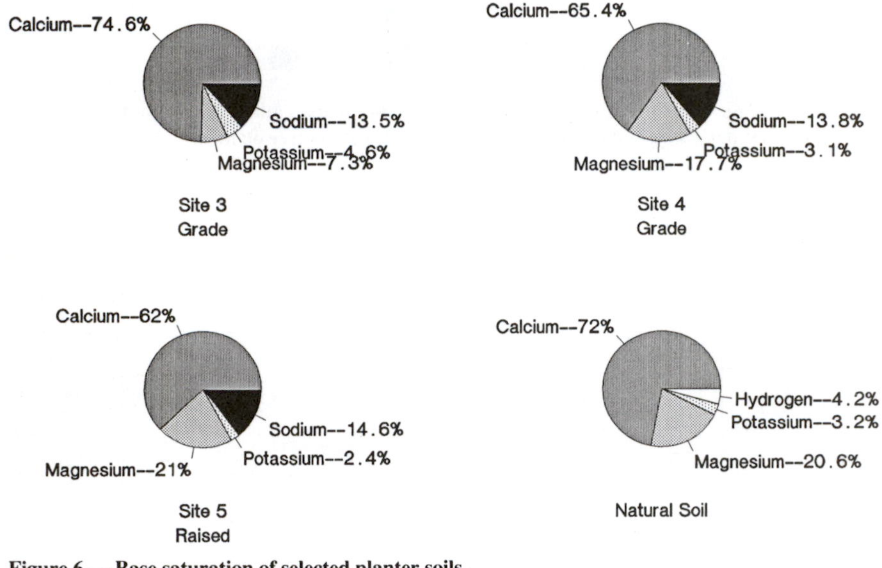

Figure 6 — Base saturation of selected planter soils.

ODRs above 0.3 μg/cm²/min. This may in part be due the underlying materials which were lithologically similar to the coarse planter soil materials. This appears to be a textural relationship and warrants further study. Oxygen diffusion rates in soils show no relationship to planter type (Figure 7) in this study.

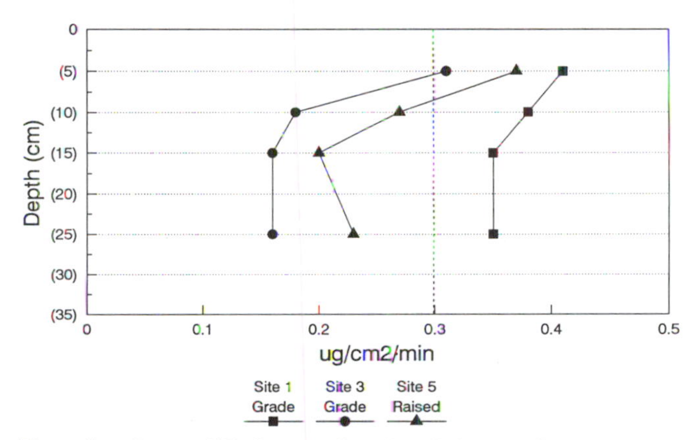

Figure 7 — Oxygen diffusion rates for selected planter soils.

Figure 8 — Two commonly used planter types in the Chicago metro region—(left) raised without drainage holes and (right) grade level with curb stop.

Planter Types

Figure 8 shows several planter types evaluated. Planter types are commonly heralded as mitigation for poor drainage or deicing salt problems (1). Raised planters have lower concentrations of sodium than grade or curb edged planters (Figure 3). This is due in part to the raised planters not being used for snow storage. Also, raised planters do not receive salt-laden snowmelt as do those at or near grade. Raised planters along high speed roadways such as Lake Shore Drive (Site 5) do not provide any reduction in salt concentrations in the soil following one winter of deicing (Figure 6). This can be attributed to the splash and spray generated by high speed traffic rising above the planter wall and deposited into the planter. The data presented in Figure 3 show large differences in sodium levels, sodium adsorption ratios and exchangeable sodium percentage among planter types. Planters that do not receive large amounts of deicing salt laden meltwater will not become sodium contaminated.

Conclusions

Planters are harsh environments in which to grow trees. Each planter type has inherent limitations which make the rooting environment hostile. Thus, the soil utilized in the planter must have optimal conditions including its chemical composition. Soil alkalinity and salinity are two major stress factors for trees in planters. Neither is easily overcome, and whole soil replacement is the preferred method of amelioration. Because of the generally unknown chemical environment in urban tree planters, it is important to test the soil for the appropriate potential limiting factors. Before planting and during planter usage, soils should be tested for organic matter content, pH, electrolytic conductivity, sodium content, base saturation, and cation exchange capacity. Exchangeable sodium and sodium absorption ratio should be calculated from the soil test results. Planter soil can only be evaluated properly for potential chemical stresses by testing for all limiting factors.

Literature Cited

1. Craul, P.J. 1992. Urban Soil in Landscape Design. John Wiley and Sons, New York, NY.
2. Glinski, J. and W. Stepinewski. 1985. Soil Aeration and Its Role for Plants. CRC Press, Boca Raton, FL.
3. Kelsey, P.D. 1990. Salt in fertilizer. Tree Care Industry 1: 18.
4. Kelsey, P.D. 1990. Report on the decline of Redmond' lindens on the property of Continental Bank, Division and Clark, Chicago, Illinois. Unpublished report of the Morton Arboretum Soil Characterization Laboratory. 14pp.
5. Kelsey, P.D. and R.G. Hootman. 1988. Soil resource evaluation for a group of sidewalk street tree planters. J. Arboric. 16:113-117.
6. Kopinga, J. 1991. The effects of restricted volumes of soil on the growth and development of street trees. J. Arboric. 17:57-63.
7. Lindsey, P. and N. Bassuk. 1991. Specifying soil volumes to meet the water needs of mature urban street trees and trees in containers. J. Arboric. 17:141-149.
8. MacDonald, J.D., L.R. Costello, and T. Berger. 1993. An evaluation of soil aeration status around healthy and declining oaks in an urban environment in California. J. Arboric. 19:209-219.
9. Messenger, S. 1984. Treatment of chlorotic oaks and red maples by soil acidification. J. Arboric. 10:122-128.
10. Messenger, S. 1986. Alkaline runoff, soil pH, and white oak manganese deficiency. Tree Physiology 2:317-325.
11. Page, A.L., R.H. Miller, and D.R. Keeney. 1982. Methods of Soil Analysis. Part 2. Chemical and Microbiological Properties. Am. Soc. Agron., Madison, WI.
12. USDA. 1954. Diagnosis and improvement of saline and alkali soils. Agric. Hndbk. 60.

Root-Knot Nematodes on Landscape Trees

Frank S. Santamour, Jr.

Many landscape trees are susceptible to infestation by common and widespread root-knot nematodes. A few nematodes with restricted host specificity have been found on several tree species and it is likely that more will be found in the future. The direct effects of nematodes on tree growth may not be readily apparent but the synergism between nematodes and fungi such as *Fusarium* and *Verticillium* can result in the death of trees considered resistant or tolerant to the wilt pathogens. Arborists, horticulturists, and nurserymen should consider at least a cursory examination of root systems when investigating the causes of poor plant performance.

Until rather recently, reports of root-knot nematodes (*Meloidogyne* spp.) on broad-leaved tree species of forests and landscapes had been largely limited to the finding of root galls on plants during quarantine inspections or nursery surveys. More often than not, such reports failed to identify the species of nematode causing the root galls and did not mention whether the nematode was reproducing on the host plant. The four most common root-knot nematodes in the Northern Hemisphere (*M. arenaria, M. hapla, M. incognita, M. javanica*) have a broad host range among crop plants. Some trees growing naturally or under cultivation were found to be susceptible to one or more of these nematodes, with obvious above-ground symptoms of infestation. One notable example of such a finding was the wilting and stunted growth of seedlings of tuliptree (*Liriodendron tulipifera*) in Georgia caused by parasitization by *M. javanica*. Inoculation studies indicated, however, that tuliptree was not susceptible to the other three common nematode species (7).

When new, and apparently more specialized, nematode species were found on tree roots, researchers generally attempted to determine the host range of these species by inoculating seedlings of a number of other woody plants. Thus, *M. ovalis*, found on roots of sugar maple (*Acer saccharum*) and white ash (*Fraxinus americana*) in the natural forests of Wisconsin, was able to infest the roots of three other maple species but not those of green ash (6). Artificial inoculation of American chestnut (*Castanea dentata*) with *M. querciana*, isolated from pin oak, showed that chestnuts, as well as oaks, might be susceptible to the oak root-knot nematode (2). The sycamore root-knot nematode (*M. platani*) was capable of parasitizing roots of white ash but not those of flowering dogwood (1,3). All of this information, and more, is contained in a comprehensive

---

Frank S. Santamour, Jr. is a Research Geneticist with the U.S. National Arboretum, Agricultural Research Service, U.S. Department of Agriculture, Washington, D.C. 20002.

compilation of reports of identified or unidentified root-knot nematodes on more than 120 species in 56 genera of landscape trees (12).

Nematodes on Willows

The blow-down of a tree of *Salix alba* during a mild overnight windstorm in 1988 stimulated our interest in root-knot nematodes. Examination of the exposed roots indicated extensive infestation by nematodes on one side of the tree that resulted in severe stunting of the root system and a loss of "anchoring" ability of the roots on the infested side. The nematode isolated from that willow was subsequently identified as *M. arenaria*-Race 2. Our first major generic survey of the susceptibility of tree species to root-knot nematodes involved the artificial inoculation of rooted cuttings of 48 willow clones with *M. arenaria* (two races), *M. hapla, M. incognita,* and *M. javanica*. The 48 clones included 25 "weeping" willows then growing under various species or cultivar names in the collections of four major arboreta, five clones/cultivars of *S. alba*, two cultivars of *S. babylonica*, and 16 other clones representing 12 additional botanical sections of two subgenera of *Salix* (13).

The "true" weeping willow (*S. babylonica* 'Babylon') was susceptible to all of the nematodes except *M. hapla*, as were the 25 other weeping, possibly hybrid, clones from various arboretum collections. The European *S. alba*, which may have been the other species parent of these weeping selections (8), was similarly susceptible. The only tree-type willow on which the nematodes did not reproduce was the American black willow (*S. nigra*). Differences in chromosome numbers between *S. nigra* and the weeping types would probably be a deterrent to hybridizations designed to develop new nematode-resistant weeping cultivars.

Nematodes on Oaks

Our survey of oaks encompassed 32 species, including representatives of the five subgenera of *Quercus* (9). Relatively few taxa, chiefly the Asiatic and European species of subgenus *Cerris*, were susceptible to any of the four common nematode species. However, all nine species of American red oaks tested, as well as one or more species in the other subgenera, were susceptible to parasitization by the oak root-knot nematode (*M. querciana*). Thus, despite the fact that *M. querciana* has been found only once (1965, on pin oak in a Virginia nursery), the wide potential host range suggests that this nematode may be far more common.

Nematodes on Maples

Eighteen species of maples (*Acer*), representing 11 botanical sections of the genus, were surveyed for susceptibility to the four common nematode species and the oak root-knot nematode (10). The maple species varied widely in their response to nematode inoculation, with some species even being susceptible to *M. hapla*, which did not parasitize any of the willows or oaks. None of the nematodes was capable of reproduction on sugar maple (*A. saccharum*) or sycamore maple (*A. pseudoplatanus*). However, red maple (*A. rubrum*), silver maple (*A. saccharinum*), and Norway maple (*A. platanoides*) were susceptible to some common nematode species. There was a high degree of variation in nematode susceptibility among the seedlings of Norway maple derived from a single tree. Such variability was also noted among four own-rooted cultivars of this species (11). As might be expected, none of the 10 maple species inoculated with the oak root-knot nematode was susceptible. Further studies on maples should be undertaken when the maple root-knot nematode (*M. ovalis*) is once again isolated and cultured.

Nematodes on Various Trees

Rather than continue host generic surveys, our most recent study examined the nematode susceptibility of 23 species in 17 genera of landscape trees, including many widely planted species (14). Many urban trees, including *Ailanthus altissima, Gleditsia triacanthos, Liquidambar styraciflua, Magnolia grandiflora, and Pyrus calleryana*, were resistant to all the common root-knot nematodes. On the other hand, *Ginkgo biloba, Koelreuteria paniculata, Robinia pseudoacacia, Sophora japonica, Ulmus parvifolia, Zelkova serrata*, and some species of *Betula* and *Prunus* were susceptible to one or more of the common nematodes.

Nematode-fungus Synergism

As noted earlier, we found variation in nematode susceptibility among own-rooted cultivars of Norway maple (11). The cultivar 'Jade Glen', which was considered highly tolerant of Verticillium wilt (15), was susceptible to three common root-knot nematode species. Young plants of this cultivar exhibited severe wilting when subjected to artificial inoculation of both the wilt fungus and the nematodes. None of the nematodes was able to reproduce on the roots of another wilt-tolerant cultivar ('Parkway'), and no plants wilted. The wilt-susceptible cultivars, 'Crimson King' and 'Greenlace', did show wilt symptoms, but they were highly resistant to the nematodes. Thus, the synergistic action of the fungus-nematode combination is not necessary for wilt to occur in Norway maples, but it has been shown that some selected wilt-tolerant trees might exhibit severe wilt symptoms if they are successfully attacked by nematodes.

Although the aforementioned study on Verticillium wilt-nematode synergism in Norway maple is a first report of this situation in trees, it has been well known that such synergism can occur between Fusarium wilt fungi and root-knot nematodes in crop plants and trees. McArdle and Santamour (4) screened 6,000 seedlings of *Albizia julibrissin*, derived from inter-pollination among trees selected for resistance to Fusarium wilt, with a combination of the wilt fungus and two root-knot nematodes. Of 78 plants that survived this intensive screening, only 20 were symptomless after eight years of field testing. The best of these are now being propagated for wider testing and eventual release as superior disease-resistant cultivars.

Current Studies

Although we have more or less abandoned the large-scale generic surveys of nematode susceptibility reported here, there are special situations that may demand such research. One such situation is the recent resumption of breeding work at the U.S. National Arboretum to develop superior disease-resistant cultivars of *Cercis*. We are currently evaluating the nematode susceptibility of several species of *Cercis* and certain provenances of *C. canadensis* as part of this program.

Preliminary work with some of the nematode-resistant tree species found earlier (14) indicated that the roots of certain trees might be producing nematicidal substances. An experiment was set up in 1993 to examine this possibility. Another study is underway to determine the potential inheritance of nematode resistance in *Albizia*.

Importance of Root-knot Nematodes

At the present time, considering our lack of knowledge, it would not be prudent to overestimate the deleterious effects of root-knot nematodes on landscape trees. We do know, however, that combinations of root-knot nematodes and certain wilt pathogens may cause considerable damage to trees. It is possible that there may also be synergistic situations, similar to those reported by Powell (5), with important root-rot pathogens such as Armillaria. Arborists and other professionals who deal with the planting, maintenance, and care of trees in the landscape should be aware that root-knot nematodes are potentially important pests and that the examination of tree root systems should be an integral part of their diagnoses of tree problems.

Literature Cited

1. Al-Hazmi, A. S., and J. N. Sasser. 1982. Biology of Meloidogyne platani Hirschman parasitic on sycamore, Platanus occidentalis. J. Nematol. 14: 154-161.
2. Golden, A. M. 1979. Descriptions of Meloidogyne camelliae n. sp. and M. querciana n. sp. (Nematoda: Meloidogynidae) with SEM and host range observations. J. Nematol. 11: 175-189.
3. Hirschman, H. 1982. Meloidogyne platani n. sp. (Meloidogynidae), a root-knot nematode parasitizing American sycamore. J. Nematol. 14: 84-95.
4. McArdle, A. J., and F. S. Santamour, Jr. 1986. Screening mimosa (Albizia julibrissin) seedlings for resistance to nematodes and Fusarium wilt. Plant Dis. 70: 249-251.
5. Powell, N. T. 1971. Interactions between nematodes and fungi in disease complexes. Ann. Rev. Phytopath. 9: 253-274.
6. Riffle, J. W., and J. E. Kuntz. 1967. Pathogenicity and host range of Meloidogyne ovalis. Phytopathology 57: 104-107.
7. Ruehle, J. L. 1971. Nematodes parasitic on forest trees. III. Reproduction on selected hardwoods. J. Nematol. 3: 170-173.
8. Santamour, F. S., Jr. 1992. Biochemical verification of hybridity in weeping willow. J. Arboric. 18: 6-9.
9. Santamour, F. S., Jr. 1992. Susceptibility of oaks to root-knot nematodes. J. Arboric. 18: 216-219.
10. Santamour, F. S., Jr. 1992. Susceptibility of maples to root-knot nematodes. J. Arboric. 18: 262-265.
11. Santamour, F. S., Jr. 1992. Influence of root-knot nematodes on Verticillium wilt of maples. J. Arboric. 18: 298-301.
12. Santamour, F. S., Jr., and J. M. Batzli. 1990. Host checklist of root-knot nematodes on broad-leaved landscape trees. J. Arboric. 16: 162-168.
13. Santamour, F. S., Jr., and J. M. Batzli. 1990. Root-knot nematodes on willows: Screening of Salix species, cultivars, and hybrids for resistance. J. Arboric. 16: 190-196.
14. Santamour, F. S., Jr., and L. G. H. Riedel. 1993. Susceptibility of various landscape trees to root-knot nematodes. J. Arboric. 19: 257-259.
15. Townsend, A. M., L. R. Schreiber, T. J. Hall, and S. E. Bentz. 1990. Variation in response of Norway maple cultivars to Verticillium dahliae. Plant Dis. 74: 44-46.